T0141110

HIMALAYAN
FERMENTED
FOODS
Microbiology, Nutrition,
and Ethnic Values

HIMALAYAN FERMENTED FOODS

Microbiology, Nutrition, and Ethnic Values

Jyoti Prakash Tamang

CRC Press
Taylor & Francis Group
Boca Raton London New York

CRC Press is an imprint of the
Taylor & Francis Group, an **informa** business

CRC Press
Taylor & Francis Group
6000 Broken Sound Parkway NW, Suite 300
Boca Raton, FL 33487-2742

First issued in paperback 2019

© 2010 by Taylor & Francis Group
CRC Press is an imprint of Taylor & Francis Group, an Informa business

No claim to original U.S. Government works

ISBN-13: 978-1-4200-9324-7 (hbk)
ISBN-13: 978-0-367-38531-6 (pbk)

This book contains information obtained from authentic and highly regarded sources. Reasonable efforts have been made to publish reliable data and information, but the author and publisher cannot assume responsibility for the validity of all materials or the consequences of their use. The authors and publishers have attempted to trace the copyright holders of all material reproduced in this publication and apologize to copyright holders if permission to publish in this form has not been obtained. If any copyright material has not been acknowledged please write and let us know so we may rectify in any future reprint.

Except as permitted under U.S. Copyright Law, no part of this book may be reprinted, reproduced, transmitted, or utilized in any form by any electronic, mechanical, or other means, now known or hereafter invented, including photocopying, microfilming, and recording, or in any information storage or retrieval system, without written permission from the publishers.

For permission to photocopy or use material electronically from this work, please access www.copyright. com (http://www.copyright.com/) or contact the Copyright Clearance Center, Inc. (CCC), 222 Rosewood Drive, Danvers, MA 01923, 978-750-8400. CCC is a not-for-profit organization that provides licenses and registration for a variety of users. For organizations that have been granted a photocopy license by the CCC, a separate system of payment has been arranged.

Trademark Notice: Product or corporate names may be trademarks or registered trademarks, and are used only for identification and explanation without intent to infringe.

Library of Congress Cataloging-in-Publication Data

Tamang, Jyoti Prakash.
    Himalayan fermented foods : microbiology, nutrition, and ethnic values / Jyoti Prakash Tamang.
        p. ; cm.
    "A CRC title."
    Includes bibliographical references and index.
    ISBN-13: 978-1-4200-9324-7 (hardcover : alk. paper)
    ISBN-10: 1-4200-9324-X (hardcover : alk. paper)
    1. Fermented foods--Himalaya Mountains Region. I. Title.
    [DNLM: 1. Food Microbiology--Asia. 2. Fermentation--Asia. QW 85 T153h 2010]

TP371.44.T36 2010
641.095496--dc22   2009011498

Visit the Taylor & Francis Web site at
http://www.taylorandfrancis.com

and the CRC Press Web site at
http://www.crcpress.com

Contents

Preface

The Himalayan region comprises a diversity of hot spots with rich biore-sources of flora and fauna. The Himalayas are also a sacred place for mil-lions of Buddhist and Hindu ethnic people. The approximate population of the Himalayas has been estimated at more than 65 million, with more than 300 major ethnic groups. This mix of ancient cultures, ethnic diver-sity, and biological resources has produced a remarkably diverse food culture, comprising a range of fermented and nonfermented ethnic foods and alcoholic beverages.

Food symbolizes the culture of a community, providing information reflecting its eating habits, consumption patterns, food preferences, nutri-tional security and community health, agricultural and livestock sys-tems, marketing strategies, socioeconomy, ethnicity, and religious taboos. More than 150 different types of ethnic fermented foods and alcoholic beverages are prepared and consumed by the Himalayan people, which include milk, vegetable, bamboo, soybean, meat, fish, cereal, and alco-holic beverages. The food culture of the Himalayas is a unique fusion of the soybean–alcohol-consuming Chinese culture from the north and the milk–vegetable eating Hindu culture from the south.

Diverse microorganisms, ranging from mycelial fungi to enzyme- and alcohol-producing yeasts and bacteria, are associated with the fermenta-tion and production of ethnic Himalayan foods and alcoholic drinks. Most of the foods are fermented naturally, except for the alcoholic beverages, which are produced by using a consortium of microorganisms in the form of dry, cereal-based starter materials. Diversity within the species of lactic acid bacteria and bacilli has created ethnic foods with different sensory characteristics. The functional microorganisms present in the Himalayan-fermented foods have many biological functions that provide a number of important benefits, including health-promoting benefits, biopreservation of perishable foods, bioenrichment of nutritional value, protective proper-ties, and therapeutic values.

Himalayan fermented foods comprise all types of available sub-strates, ranging from milk to alcohol, soybeans to cereals, vegetables to bamboo, meat to fish, and alcoholic beverages to nonalcoholic beverages.

The historical record of consumption of milk and milk products in Nepal in 900 b.c. throws light on the cultural history of food habits of the Himalayan people. Although the diversity of the Himalayan foods is unknown to most of the countries outside the Himalayas, these ethnic foods have been consumed by the Himalayan people for more than 2500 years.

This book has ten chapters covering the indigenous methods of preparation, culinary practices, socioeconomic impacts, microbiology, functional properties, nutritional values, antiquity, and ethnic values of Himalayan fermented foods. I have tried to document and update the information on the indigenous knowledge of production methods, history (antiquity), and ethnic values based on field surveys and relevant historical documents. The microbiological, biochemical, nutritional, and functional aspects are mostly based on our primary research works. Perhaps, I am one of the few scholars to research works on Himalayan fermented foods for more than two decades. Today, there are many universities and research institutes in India and abroad working on different aspects of fermented foods and beverages, including the Himalayas. Some of the researchers are my students.

I am grateful to Taylor and Francis Group for publishing this book. I hope it will be referenced by researchers, students, teachers, tourists, travelers, media persons, food entrepreneurs, agriculturalists, government policy makers, anthropologists, ethnologists, and sociologists who have an interest in the Himalayas and their people and bioresources. There are many books that address the myriad issues and aspects of the Himalayas, but a book that specifically describes the Himalayan foods and food culture is rare. This book, *Himalayan Fermented Foods: Microbiology, Nutrition, and Ethnic Values*, published by Taylor and Francis Group, is a beginning.

I salute my ancestors for providing a rich diversity of ethnic fermented foods and drinks and for maintaining the unknown microbial genetic resources with vast bionutrients and health-promoting benefits.

Jyoti Prakash Tamang
Darjeeling, India

Acknowledgments

I am thankful to my wife Dr. Namrata Thapa for constant support and technical assistance in the preparation of this book. Over the past 16 years, the team of brilliant Ph.D. students that I have recruited from the Food Microbiology Laboratory, Sikkim Government College, Gangtok, has been the real driving force in researching and identifying the scientific mechanisms of ethnic Himalayan fermented foods. I express my deepest appreciation to all my former students. Some of them are placed in good jobs elsewhere, and some are present students. The list includes Dr. Saroj Thapa, Dr. Sailendra Dewan, Dr. Hannah Yonzan, Dr. Buddhiman Tamang, Dr. Arun Rai, Mr. Rajen Chettri, Ms. Nikki Kharel, Mr. Bimal Rai, Mr. Sudhan Pradhan, Mr. Rudra Mani Sharma, Ms. Jayasree Charkraborty, and many others. I am also grateful for the efforts of the supporting staff and Mr. Karma Tamang. My sincere thanks to Mr. Satyam Thapa and Mr. Ravi for assisting me in the computations presented in this book. My special thanks to the following researchers for helping me in preparation of this book: Dr. Diwakar Pradhan of Banaras Hindu University; Dr. Vimal Khawas of Sikkim University; Dr. Toshirou Nagai of the Institute of Agrobiological Sciences, Tsukuba; Dr. A. A. Mao of the Botanical Survey of India; Mr. J. R. Subba of Gangtok; Mr. Bejay Chettri of Deokota Sangh, Siliguri; and Mr. H. K. Pradhan of Siliguri.

My sincere thanks to the various cultural associations, societies, and individuals of Darjeeling hills, Sikkim, Nepal, Bhutan, Uttarakhand, Himachal Pradesh, Jammu and Kashmir, and all North East Indian states for their cooperation in compilation of the food habits of the communities, and also to all Himalayan ethnic people, to which I belong.

I have been inspired by my beloved father (Aapa), the late Capt. Surjaman Tamang; my mother (Aama), Mrs. Prava Devi Tamang; and my father-in-law (Baba), the late Bhuwan Singh Thapa. I dedicate this book to all of them. My daughter Ashweela and my son Ashwaath missed the vacation due to my preoccupation with this book, but they realized the importance of this book and supported me. My love to them and to Namrata.

I am grateful to Taylor and Francis Group for offering me the opportunity to write this book, particularly Dr. Stephen Zollo, Ms. Jennifer Ahringer, and Ms. Gail Renard.

Jyoti Prakash Tamang

About the Author

Dr. Jyoti Prakash Tamang, a leading authority in food microbiology, has been studying the Himalayan fermented foods and beverages concerning microbiology, nutritional aspects, functionalities, and food cultures for the last 22 years. He earned a B.Sc. (honors), M.Sc., and Ph.D. from North Bengal University, followed by postdoctoral research at the National Food Research Institute (Japan) and the Institute of Hygiene and Toxicology (Germany). Dr. Tamang was awarded the Gold Medal by North Bengal University in 1985, the National Bioscience Award of Department of Biotechnology of Ministry of Science and Technology of India in 2005, and became a Fellow of the Biotech Research Society of India in 2006.

Dr. Tramang has published more than 75 research papers in peer-reviewed international and national journals, has filed a patent, and is guiding several students. He has presented his works in 17 different countries and is a member of several prestigious national and international academic and scientific organizations. He is a chief editor of *Journal of Hill Research*, and he is a regular reviewer of many reputable national and international scientific journals and books. Professor Tamang is the president of the Centre for Traditional Food Research at Darjeeling, a team leader of the Food Microbiology Laboratory, and a senior visiting faculty in microbiology at Sikkim Central University.

chapter one

The Himalayas and food culture

1.1 The Himalayas

The great Himalayas are a sacred place for millions of Buddhist and Hindu people and the center of a rich diversity of cultures and biological resources. The meaning of the word *Himalayas* in Sanskrit is "abode of snow." The Himalayan arc extends between latitudes 26°20′ and 35°40′ north and between longitudes 74°50′ and 95°40′ east (Ives 2006). The Himalayas extend from the Indus Trench below Nanga Parbat (8125 m) in the west to the Yarlungtsangpo-Brahmaputra gorge below Namche Barwa (7756 m) in the east—a west–northwest to east–southeast distance of about 2500 km and a width of 200–300 km—and include India (Jammu & Kashmir, Himachal Pradesh, Uttarakhand, Sikkim, Darjeeling hills, Arunachal Pradesh, and some hill regions of northern Assam), Nepal, Bhutan, and Tibet Autonomous Region in China. The region directly provides a life-support base for over 65 million mountain people (Khawas 2008). The Himalayas form the highest mountain region in the world with more than 30 peaks, and one of them is Mount Everest (8848 m), the highest mountain in the world. Mount Kangchendzonga (8579 m), the third-highest peak in the world, is India's highest mountain peak and rises from Sikkim.

Geographically, the Himalayan mountain system is divided into (a) the Greater Himalaya Himadri area above the main central thrust, consisting of snow-clad peaks, glaciers, and ranges of mountains, (b) the Lesser Himalaya Himachal, which is separated from the Himadri by the main central thrust in the north and by the main boundary thrust in the south, consisting of high mountains cut into deep ravines and precipitous defiles, and (c) the sub-Himalayan tract Sivalik, the foothill belts of the region, consisting of the latest geological formation of loose boulders and soil (Pradhan et al. 2004).

The Himalayas are categorized into the following divisions based on population and vegetation: western Himalayas (Jammu & Kashmir, Himachal Pradesh), central Himalayas (Uttarakhand, western Nepal), and eastern Himalayas (eastern Nepal, Darjeeling hills, Sikkim, Arunachal Pradesh, hills of North East India, Bhutan, and Tibet Autonomous Region in China).

The geographical locations of the Indian Himalayas have been described by Nandy et al. (2006). The Kashmir Himalayas occupy the geographical location between latitudes 32°17′–37°5′ north and longitudes 72°40′–80°30′ east, with a total area of 222,236 km². Kashmir has borders with Afghanistan in the northwest, with Pakistan in the west, and with China in the north. The whole region is differentiated into four broad groups—Karakoram, Ladakh, Zaskar, and Pir Panjal—and these mountain ranges are separated by deep gorges, forming the valleys of the Shyok, Indus, and Jhelum rivers, respectively. The Himachal Himalayas lie between the latitudes 30°23′–33°13′ north and longitudes 75°43′–79°4′ east, with a total area covering 55,673 km². The state of Himachal Pradesh lies to the south of Jammu & Kashmir state. The state is bound in the east by China, the Garhwal region of Uttarakhand state in the southeast, Punjab state in the southwest, and in the south by Haryana state. The Uttarakhand Himalayas are geographically located between latitudes 29°5′–31°25′ north and longitudes 77°45′–81° east, covering an area of 51,124 km². The region comprises two administrative units of Uttarakhand state, i.e., Garhwal (northwest) and Kumaun (southeast). The eastern Himalayan region lies between the latitudes 26°40′–29°30′ north and longitudes 88°5′–97°5′ east and covers a total area of 93,988 km², comprising Darjeeling hills, Sikkim, and Arunachal Pradesh. Bhutan is located between the Tibetan plateau and Assam-North Bengal plains of India, and has borders with Sikkim in the west, China (Tibetan Autonomous Region) in the north, and Arunachal Pradesh in the west. The Purvanchal Himalayas lie between the latitudes 21°5′–28°23′ north and longitudes 91°13′–97°25′ east, covering a total area of 108, 229 km², comprising the hills of Assam (15,322 km²), Manipur (22,327 km²), Meghalaya (22,429 km²), Mizoram (21,081 km²), Nagaland (16,579 km²), and Tripura (10,491 km²). The region of North East India has international boundaries in the east with Myanmar, in the south and west with Bangladesh, in the northwest with Bhutan, and in the north with Tibet Autonomous Region of China.

1.2   Agriculture in the Himalayas

The agroclimatic conditions of the Himalayas vary from hot, subhumid tropical in the southern low tracts to temperate, cold alpine, and glacial in the northern high mountains, due to their various subecological locations, elevation, and topography. The temperature varies from the lowest recorded temperature in Leh in the Ladakh region of Jammu & Kashmir in India at –28.3°C and the highest recorded temperature in Jammu in Jammu & Kashmir state of India at 47.2°C (Singh 1991).

The lower valleys and gorges are very dry, and local agriculture production depends upon snowmelt and glacial-melt irrigation, commonly called *kuhl* in Himachal Pradesh (Ives and Messerli 1989). Natural vegetation belts range from tropical monsoon rain forest or *sal* forest (*Shorea robusta*) in the south, through a series of forest belts, to the upper timberline at approximately 4000–4500 m (Ives and Messerli 1989). Above this, the *Rhododendron*-shrub belt gives out onto alpine meadows, a subrival belt of extensive bare ground and scattered shrubs, herbs, bryophytes, pteridophytes, and lichens, and finally at 5000–5500 m, permanent ice and snow with steep rock outcrops (Samant and Dhar 1997).

The ethnic people of North East India, mainly Nagaland, have adopted the traditional practice of *jhum* cultivation, wherein land is cleared of its natural vegetation and farmed until it can no longer sustain production, at which point it is abandoned and left to regenerate itself with natural vegetation (Barthakur 1981). As an agricultural system, *jhum* cultivation has the disadvantage of deteriorating ecological balances and accelerating soil erosion; however, the *jhum* system is also productive and sustainable, with multiple intercropping of up to 60 food crops in one field (Barthakur 1981).

Agriculture and livestock are the major livelihood in the Himalayas, where the ethnic people have traditionally practiced integrated agriculture, animal husbandry, agro-forestry, and forestry. Over 85% of the population is directly or indirectly dependent on agriculture for its livelihood (Khawas 2008). Mountain geography and inaccessibility due to difficult terrain and lack of infrastructure have compelled the people to adopt the agro-biodiversity system, although commercial agriculture is not as high yielding and profitable as in the plains areas (Nandy and Samal 2005). Forest coverage in the Indian Himalayas is over 52% of the total reported area, followed by wastelands and agricultural land (Nandy et al. 2006).

The varied topographic and agroclimatic conditions permit the cultivation of a wide variety of indigenous crops and fruits, ranging from subtropical to cool temperate. The Himalaya region has been a rich genetic source of many indigenous varieties of agricultural plants: cereals such as rice, maize, finger millet, wheat, buckwheat, barley, sorghum, pearl millet; pulses such as soybeans, black gram, green gram, garden peas, black lentils, French beans; vegetables such as cabbage, cauliflower, leafy mustard (*rayo sag*), young tendrils and fruits and tubers of squash (*iskus*), brinjal, chilli, cucumber, young tendrils and fruits of pumpkin, sponge gourd, tomato, tree tomato, lemon, etc.; tubers and rhizomes such as potato, beetroot, sweet potato, cassava, arum/taro, yam, ginger, turmeric, large cardamom; and roots such as radish, carrot, etc. A wide variety of seasonal fruits such as orange, apple, mango, papaya, guava, banana, pear, peach, fig, avocado, etc., are also cultivated and consumed (Annual Progress Report 2005; Nandy and Samal 2005).

Tea, ginger, large cardamom, garlic, medicinal and aromatic plants, wild and domesticated ornamental plants, and orchids are the cash generators for the people. Bee-farming (both domestic and wild bees) for honey is also a common practice among the Himalayan farmers. Some indigenous varieties of chilli—locally called *dalley khorsani* (round red/green chilli) of Darjeeling hills, Sikkim, and Nepal; *uamorok* of Manipur; and *raja* of Nagaland—are among the hottest chillies in the world and are promising agricultural products in the Himalayan regions. Varieties of wild edible plants, including young bamboo shoots, ferns, stinging nettles, and their parts such as seeds, fruits, roots, leaves, and flowers are part of the local diet eaten by the Himalayan people (Sundriyal and Rai 1996; Sundriyal and Sundriyal 2004; Rai et al. 2005).

More than 78 indigenous and exotic species of bamboo belonging to 19 genera are found in the biodiversity-rich regions of North East India (Hore 1998). About 26.2 tons, 435 tons, and 426.8 tons of bamboo shoots are harvested annually in Sikkim, Meghalaya, and Mizoram states, respectively, located in the North East region of India (Bhatt et al. 2003).

Due to the predominant agrarian economy in the Himalayas, animal husbandry plays a vital role for supporting agricultural operations to supplement food and as sources of manure. The domestic livestock of the Himalayas includes cow, ox, goat, pig, sheep, yak, joe/churru (hybrid of cow and yak), buffalo, poultry, etc., which are mainly used for meat, milk and milk products, and eggs. Yaks (*Bos grunniens*) are reared mostly on extensive alpine and subalpine scrublands between 2100 and 4500 m in altitude for milk products, meat, hairs, tails, and skins (Pal et al. 1995; Balaraman and Golay 1991; Sharma et al. 2006).

The river systems along with their tributaries in the Himalayas exhibit a wide range of gradients from subtropical to alpine zone. Many indigenous species of fish are found in the rivers of Sikkim and Darjeeling hills (Thapa 2002). The Brahmaputra and its tributary rivers in Assam and Arunachal Pradesh consist of more than 126 species of fish belonging to 26 families (Motwani et al. 1962; Jhingran 1977). Logtak lake in Manipur, which provides the main fishery resources in Manipur and other adjoining states, has varieties of ichthyofauna mostly dominated by species of *Puntius, Channa, Anabas,* etc. (Chaudhuri and Banerjee 1965). The people of North East India catch the available fish from various sources, mainly from rivers, streams, and lakes, and consume fresh as well as traditionally processed fish products (Thapa et al. 2006). Inland fishery programs in the lower altitudes of some of the Himalayan regions are becoming popular among farmers as a source of income.

1.3   Ethnic people

The Hindu epics in the history of India identify the original inhabitants of the Himalayas as the Kinnar, Kilind, and Kirat (O'Flaherty 1975; Ives and Messerli 1989). The Negroids, the Mongoloids, and the Aryans form the macro social groups of the Himalayan population (Ives and Messerli 1989; Khawas 2008).

The Hindus of Indian origin mainly dominate the sub-Himalayan and middle Himalayan valleys, while the Great Himalayan region is influenced by the Tibetan Buddhists (Khawas 2008). The ethnicity of the Gilgit, Baltistan, and Poonch regions of Kashmir are overwhelmingly Muslim; Jammu is mainly Hindu; and Ladakh is predominantly Buddhist (Tibeto-Mongoloids). The semiarid highland zone of Himachal Pradesh, the trans-Himalayan tracts of Lahaul-Spiti and Kinnaur, are inhabited by Tibetans, while the other parts of Himachal Pradesh are mainly inhabited by the Hindus. Culturally, Uttarakhand is largely dominated by the Hindu Kumauni and the Garhwali culture in the middle and low altitudes, while in the northern high-altitude valleys, the Bhotia or Bhutia of the Tibetan origin predominates. Nepal has a blend of both Hindu and Buddhist cultures, producing a mixed culture of Indian and Tibetan traits. Darjeeling hills and Sikkim is a mixture of both Hindu and Tibetan culture, while Bhutan has historically been a region of Tibetan culture. Arunachal Pradesh reflects the religion and culture similarity with the Chinese of the Yunnan province in China, while the northern part of Arunachal Pradesh has predominantly the races of Tibetan origin. North East Indian states bordering Myanmar have cultural and social affinities with the people of Myanmar. Further, notable proportions of Christians also live in Meghalaya, Mizoram, Nagaland, Darjeeling hills, Sikkim, Arunachal Pradesh, and Nepal. Migrants of Muslim population have also been observed recently in the demography of the Himalayas except Kashmir, which is the predominant ethnic Muslim region. The approximate population of the Himalayas has been estimated to be more than 65 million (Khawas 2008), with more than 171 major ethnic communities living in the Indian Himalayas (Samal et al. 2000) and 61 ethnic groups in Nepal (C. Subba 1999).

The major ethnic groups living in the Himalayas are summarized as follows based on data compiled from the reports of Jamir and Rao (1990), Census of India (2001), and Nandy et al. (2006) as well as my personal collection: Jammu & Kashmir (Dogra; Gujjar; Gaddi; Kashmiri Pundit; Sunni, Shia, Hanji, and Dard Muslims; Balti; Ladakhi), Himachal Pradesh (Rajput, Brahmin, Ghairat, Mahajan, Sood, Chahang, Saini, Air, Darni, Lohar, Tarkhan, Nai, Dusali, Doomna, Chamar, Julaha), Uttarakhand (Kol or Kolta, Rajput, Brahmin, Jaunsari, Bhotia, Buksha, Tharu), Darjeeling hills (ethnic Gorkha/Nepali [Rai, Tamang, Gurung, Limboo, Chettri, Magar, Bahun, Pradhan/Newar, Dewan, Sunwar, Bhujel, Khagatey, Sherpa, Sanyasi/Giri, Kami, Damai, Sarki, Maji], Lepcha, Tibetan), Sikkim (ethnic Nepali [Bahun, Chettri, Sanyasi/Giri, Magar, Tamang, Pradhan/Newar, Rai, Limboo, Gurung, Bhujel, Dewan, Sunwar, Khagatey, Sherpa, Kami, Damai, Sarki, Maji], Lepcha, Bhutia), Arunachal Pradesh (Monpa, Sherdukpen, Memba, Khamba, Khampti, Singpho, Adi, Aka, Apatani, Bangni, Nishing, Mishmi, Miji, Tangsa, Nocte, Wancho), Assam (Ahom, Bodo, Karbi, ethnic Nepali,

Miri, Rabha, Bengali), Meghalaya (Khasi, Garo, Jaintia, ethnic Nepali/ Gorkha), Mizoram (Mizo, Hmar, ethnic Gorkha, Lakher, Pawi), Manipur (Meiti, Kuki, ethnic Gorkha), Nagaland (Angami, Chakhesang, Ao, Sema, Rengma, Lotha, Chang, Konyak, Sangtam, Phom, Zeliang, Mao, Maram, Tangkhul, Maring, Anal, Mayao-Monsang, Lamkang, Nockte, Haimi, Htangun, Ranpan, Kolyo, Kenyu, Kacha, Yachimi, Kabui, Uchongpok, Makaoro, Jeru, Somra, ethnic Gorkha), Tripura (Chakmas, Bengali), Nepal (Newar, Magar, Tamang, Rai, Limboo, Gurung, Bahun, Chettri, Dewan, Sanyasi, Bhujel, Sunwar, Khagatey, Sherpa, Kami, Damai, Sarki, Yadav, Taru, Mahji, Kumhal, Urau, Meche, Dhimal, Satar, Rajbanshi), Bhutan (Drukpa/Ngalop, Sharchop, and ethnic Nepali/Lhotshamp), Tibetan Autonomous Region (Tibetan, Chinese).

Topographically, linguistically, and culturally, the ethnic Nepali or Gorkha of Nepal, Darjeeling hills, Sikkim, and south Bhutan have similarities, with more than 20 major castes within the community (Tamang 1982). Present-day India has a sizable number of ethnic Nepali or Gorkha (Gorkha denotes the distinct ethnic Nepali of Indian origins and differentiates from the citizens of Nepal by an official gazette notification of the Indian Ministry of Home Affairs in 1988) in many states of India who contribute to the major food culture of the regions. The Gorkha inhabitants in India reside in Darjeeling hills, Sikkim, a few parts of North East India, Uttarakhand, and Himachal Pradesh. Bhutan has two distinct ethnic communities, the Drukpa of Buddhist origins and the ethnic Nepali of both Hindu and Buddhist origins. The Nepali or Gorkha typically lives a pastoral and agrarian lifestyle and has mixed culture of Aryan and Mongoloid traits.

1.4   Food culture

Food symbolizes the culture of a community and provides it with a distinct identity. The social forms, customary beliefs, and material traits of racial, religious, or social groups are some of the characteristics contributing to the description of a culture, while ethnicity is the affiliation with a race, people, or cultural group (McWilliams 2007). Culture and ethnicity are essential foundations of the study of food and people. Religion is a strong factor in cultural identity, and the shared common beliefs and practices that are central to a particular religion create common threads that bind people together into a culture (McWilliams 2007). Hindu food follows the concept of purity and pollution, which determines interpersonal and intercaste relationships (Misra 1986). North Indian Brahmin kitchens produce two types of meals: *kaccha*, which means unripe and uncooked, and *pakka*, meaning ripe and cooked. *Kaccha* foods are highly vulnerable to contamination and, therefore, there are strict codes of cooking, serving, and eating this food. *Pakka* foods are fried, so they are not as vulnerable (Misra 1986).

Ethnic foods are generally categorized into fermented foods, including alcoholic beverages, and nonfermented foods. These inexpensive and culturally accepted ethnic foods provide the basic diet for the people. Ever since ethnic people have inhabited the Himalayan regions—ranging from the foothills to the alpine region—hunting, gathering, and utilization of available plants, animals, and their products for consumption started and gradually emerged as the ethnic food culture of the present day (Tamang 2001a). Ethnic food culture harnesses the cultural history of ethnic communities, including their indigenous knowledge of food production with vast nutritious qualities as well as microbial diversity associated with fermented foods, as a genetic resource. The food culture or dietary culture of the Himalayan people presents a kaleidoscopic panorama (Tamang 2005a). Each food prepared by different ethnic communities in the Himalayas is unique and unparalleled, due to the geographical location, food preferences, and availability of raw substrates or locally grown agricultural products or animal sources.

The Himalayan culture is wedged between the rich Hindu-Aryan culture in the south and the Buddhist-Mongolian culture in the north. Thus, the Himalayan food culture is a fusion of the Hindu and the Tibetan cuisines, with modifications based on ethnic preference and social ethos over a period of time. The Himalayan regions bordering Tibet (in China)—Ladakh, Lahul & Spiti, Kinnaur & Kalpa, Chamoli, Pithorgarh, hills of Nepal, Sikkim, Darjeeling hills, Bhutan, and north Arunachal Pradesh—have a close cultural affinity with the Tibetans, which has influenced the food cultures of the regions, while in the predominantly Hindu regions, mostly in the foothills and *tarai* or plain areas of the Himalayas, have been influenced by the vegetarian diets of the Hindus. Migration of people carrying the food culture and habits may also influence the settlers, thus leading to cultural amalgamation or fusion over a period of time. Moreover, the political history of Nepal has undergone several changes in its demography since 600 b.c. until the eighteenth century a.d. (Pradhan 1982). The greater Nepal extended from Himachal Pradesh to the Far East of Assam until 1816, when the British annexed the territories of Nepal into India by signing the famous Treaty of Sugauli on 4 March 1816 (Pradhan 1982). Darjeeling hills was a territory of the Sikkim kingdom until 1835, when it was handed over to the British on a lease agreement by the *chogyals* or kings of Sikkim (Bhanja 1993).

The Himalayan ethnic foods have evolved as a result of traditional wisdom and the experiences of generations over a period of time, based on agroclimatic conditions, availability of edible resources, ethnic preference, customary beliefs, religion, socioeconomy, regional politics, cultural practices, and taboos or social bans imposed by different rulers from time to time. Rice or maize is a staple food in the eastern Himalayas, whereas wheat or barley is a staple food in the western Himalayas.

Bhat-dal-sabji-tarkari-dahi/mohi-achar combination, which corresponds to steamed rice-legume soup-vegetable-curry-curd/buttermilk-pickle, is a typical recipe of every meal in the eastern Himalayas, and roti/chapati-*dal-sabji-dahi-achar*, corresponding to baked bread/roti-legume soup-vegetable curry-curd-pickle, is a typical recipe of every meal in the western Himalayas. Though the people of the eastern Himalayas are mainly rice eaters, nowadays, roti or chapati (wheat-based baked bread) is replacing this traditional habit, particularly among the urban population. *Dhenroh* (boiled maize flour) is substituted for rice and is commonly eaten with *mohi* (buttermilk) in rural areas in Nepal, Bhutan, Darjeeling hills, and Sikkim. The food culture of the Himalayas is very rich, having more than 100 types of ethnic fermented food and 50 types of ethnic alcoholic beverages, more than 300 types of nonfermented ethnic foods, and about 350 wild edible plants as staples, snacks, condiments, refreshments, desserts, pickles, alcoholic drinks, savories, etc.

The daily life of a typical Himalayan person (in this case a typical example of a Nepali of the Himalayas) starts in the morning with a full mug of tea taken with sugar or salt, with or without milk and with or without a pinch of black pepper. The first meal eaten in the morning is a simple recipe containing *bhat-dal-sabji-tarkari-dahi/mohi-achar*. *Tarkari* meaning side dish or curry, includes a variety of ethnic fermented and nonfermented food items. It is followed by light refreshment with mostly traditional snacks and tea in the afternoon. The second major meal is dinner, which is served early in the evening, and which consists of the same *bhat-dal-sabji-tarkari-dahi/mohi-achar*. The food culture of *matwali* Nepali (who drink alcohol as a part of their social provision) includes ethnic fermented beverages and distilled alcoholic drinks as part of the evening meal (Tamang 2009). Tibetans, Bhutia, Drukpa, and Lepcha usually eat *tukpa* (noodles in soup), *skiu* or *momo* (small dumplings of wheat flour with meats), baked potatoes, *tsampa* (ground roasted barley grains), *chhurpi* (cottage cheese), *kargyong* and *gyuma* (sausages), butter tea, and *chyang* (alcoholic beverage). The ethnic people of North East India have a social provision of drinking traditional alcohol with fermented or smoked fish and other dishes. In the high mountains (above 2500-m elevation), yak milk and its products are popular food items. Milk and milk products are more popular in the western Himalayas than the eastern Himalayas except for Darjeeling hills, Sikkim, and Bhutan. In North East India, except for Assam and Tripura, milk and milk products are not the traditional food items; hence, fermented milk products have not been reported from Meghalaya, Nagaland, Mizoram, Manipur, and several parts in Arunachal Pradesh, where pastoral systems are rare.

The common connotation in Nepali cuisine is *chawrasi vyanjanas*, i.e., Nepali cuisine has 84 different food items. The Himalayan food is less spicy and is traditionally prepared in butter from cow milk or yak milk, although commercial edible oil is now also used. The majority of the Himalayan ethnic people are nonvegetarians except for the Brahmin communities, who are strict vegetarians. Due to health-conscious and rapid electronic advertisements on health and nutrition, the food habits in the urban areas of the Himalayas are changing to the vegetarian diet. Nonvegetarians eat chicken, eggs, mutton, lamb, chevon (goat), pork, beef, buffalo, yak, fish, etc. Beef and yak are taboo to a majority of the communities belonging to the Nepali, Garwhali, Kumauni, Assamese, and people from Himachal Pradesh. The Newar/Pradhan people prefer to eat buffalo meat. Cooking is usually done by daughters-in-law, daughters, and mothers. The custom of serving meals to the elder male members in the family is prevalent in the Himalayan food culture. Daughters-in-law and daughters eat afterwards, a tradition that is still followed in the rural areas. This trend of serving meals to elders and eating separately is changing in the urban areas, where families often eat together.

The Himalayan peoples have unique practices of storing of foods. *Maa* (in the Tibetan language) butter made from yak milk is commonly stored in the animal's stomach in the Tibetan villages in Ladakh (Attenborough et al. 1994), whereas *gheu* (in the Nepali language) butter made from cow milk is stored in a wooden or bamboo container by the Nepali in Nepal, Darjeeling hills, and Sikkim. This is due to the local availability of storage containers, since bamboo is not grown in alpine regions.

The Himalayan people use their hands, after washing properly, to feed themselves. Eating food by hand has been mentioned in the history of Nepal during the Lichchhavi dynasty, which spanned from 100 to 880 a.d. (Bajracharya and Shrestha 1973). Bamboo chopsticks are commonly used by the Tibetans. The use of chopsticks is not the tradition of the Himalayan people, except for the Tibetans. Plates made of brass or lined with brass, called *kasa ko thal*, are traditionally used by the majority of rural Nepali, Kumauni, Garwhali, etc. The rich people or rulers used to feed from golden or silver plates with decorations. Small soup bowls are used by the Tibetans, Bhutia, Lepcha, and Drukpa of Bhutan. Locally available tree leaves, commonly from the fig plant, are also used in some parts of North East India for serving foods.

1.5   Fusion of Western and Eastern food cultures

Rice is a staple food for millions of people in Southeast Asia, whereas wheat or barley is a staple food in Northwest Asia, Europe, America, and Australia. A typical diet of Southeast Asia comprises rice, followed by fermented and nonfermented soybean products, pickled vegetables, fish, and alcoholic beverages, whereas a typical diet of Northwest Asia comprises wheat/barley as a staple food, followed by milk and fermented milks, and meat and meat products; wine is a typical dietary culture of Northwest

Asia, Europe, America, and Australia. African food habits include both fermented and nonfermented cereal products, wild legume seeds, tubers, and cassava as staple foods, followed by meat, milk products, and alcoholic beverages. In Asia, fruits are eaten directly, whereas in Europe, America, and Australia, fruits are fermented into wine. It is interesting to note that the Himalayan food culture has both rice and wheat or barley as staple foods along with varieties of fermented and nonfermented foods prepared from soybean, vegetable, bamboo, milk, meat, fish, cereal, etc. Himalayan cuisines also include diverse types of wild edible plants and fruits. Milk and meat products, similar to the West, and fermented soybeans and alcoholic beverages, similar to the East, are common in the food culture of the eastern Himalayas. Traditionally, animal milk is not consumed by the Chinese, Koreans, Japanese, etc., despite of an abundance of milk-producing animals in their possession. On the other hand, the Indo-Europeans, Semites, and the nomadic tribesmen of North Central Asia are milk drinkers (Laufer 1914). In the Far East, the soybean, sometimes called the "cow of China," is utilized in liquid, powder, or curd forms to make soy milk, tofu (soy curd), and a number of fermented soybean products such as miso, shoyu, *natto*, tempeh, *sufu*, etc. (Hymowitz 1970).

Examination of the microbiology of traditional fermented foods of East and Southeast Asia shows that the molds (*Rhizopus, Mucor, Aspergillus*) are the predominant mycelial fungi, followed by amylolytic and alcohol-producing yeasts (*Saccharomyces, Saccharomycopsis, Pichia*), nonpathogenic species of bacilli (mostly *B. subtilis*), and few lactic acid bacteria (LAB). In contrast, in Northwest Asia, Europe, America, Australia, and Africa, fermented food products are prepared exclusively using bacteria, mostly lactic acid bacteria or a combination of lactic acid bacteria and yeasts; molds seem to be seldom and bacilli are never used in production of fermented foods except in a few ethnic African foods such as *dawadawa* and *iru* (fermented African locust bean products). However, in the eastern Himalayas, all three major groups of microorganism (mycelial fungi, alcohol- and enzyme-producing yeasts, and bacteria) are associated with ethnic fermented foods and beverages. Most of the fermented foods in Asia are prepared naturally or spontaneously, whereas in Western countries, the majority of foods are prepared by using pure starter cultures (either strains of LAB or a combination of LAB and yeasts). The use of traditionally made mixed, dry starter cultures containing filamentous molds, starch-degrading and alcohol-producing yeasts, and LAB for production of alcoholic beverages is a common practice in Southeast Asia, including the Himalayas. *Bacillus*-fermented sticky soybean foods of the eastern Himalayas are similar to sticky fermented soybean foods of other South Asian countries such as *pepok* of northern Myanmar, *thua-nao* of northern Thailand, *sieng* of Cambodia, *chungkokjang* of Korea, and *natto* of Japan. However, consumption of *Bacillus*-fermented sticky soybean food is uncommon in the rest of the world, including entire Indian states except North East India. The food culture, mainly the ethnic fermented foods, of the eastern Himalayas is probably a transition of the food cultures of the East and West.

1.6 What are ethnic fermented foods?

Ethnic fermented foods are produced, based on the indigenous knowledge of the ethnic people, from locally available raw materials of plant or animal sources either naturally or by adding starter culture(s) containing functional microorganisms that modify the substrates biochemically and organoleptically into edible products that are culturally and socially acceptable to the consumers. Functional bacteria—mostly lactic acid bacteria, yeasts, and filamentous molds—bring biotransformation of the raw or cooked materials of plant or animal origin during fermentation, enhancing nutritional value; prolonging the shelf life; improving flavor, texture, and aroma; and also exerting several health-promoting benefits (Steinkraus 1996; Hansen 2002; Tamang 2007a). Fermented foods are generally palatable, safe, and nutritious (Campbell-Platt 1994; Geisen and Holzapfel 1996; Kwon 1994). Traditional methods of food fermentation serve as affordable and manageable techniques of biopreservation of perishable agricultural or animal products without refrigeration, freezing, cold storage, and canning. The essential objective of food fermentation is to carry over supplies from the times of plenty to those of deficit.

The Himalayan people developed such innovative technology without any knowledge of food microbiology or the actual scientific mechanism involved in it, but certainly they know how to use the beneficial (functional) microorganisms for production of foods of their choice and know what needs to be done to get the desired food products. The microbiology, biochemistry, nutrition, functionality, toxicology, food safety, and biotechnology of their fermented foods are unknown to the producers, who are simply practicing an age-old food fermentation technology.

The ethnic fermented foods and beverages, which include soybeans, milk, meat, fish, vegetable, cereals, and alcoholic beverages, constitute the basic dietary culture of the Himalayan people using their indigenous knowledge of food preservation and substrate utilization (Tamang et al. 1988; Tamang 2005b, 2009). More than 150 different types of familiar and less-familiar ethnic fermented foods and beverages of the Himalayas have been listed in Table 1.1. Most of these Himalayan fermented foods and beverages have been studied by our team of researchers for the last 20 years. The history of the development of most of these ethnic fermented foods is lost; however, some of these ethnic foods have been prepared

Table 1.1 Himalayan Fermented Foods and Beverages

Food	Substrate	Nature of product	Microbes	Region
1. Fermented vegetables				
Gundruk	Leafy vegetable	Dried, sour-acidic; soup, pickle	LAB	Nepal, Darjeeling hills, Sikkim, Bhutan
Inziang-sang	Mustard leaves	Dried, sour; soup, curry	LAB	Nagaland, Manipur
Inziang-dui	Mustard leaves	Liquid, sour; condiment	LAB	Nagaland, Manipur
Goyang	Green vegetable	Freshly fermented; juice as condiment, soup	LAB	Sikkim, Nepal
Sinki	Radish tap-root	Dried, sour-acidic; soup, pickle	LAB	Nepal, Darjeeling hills, Sikkim, Bhutan
Khalpi	Cucumber	Sour; pickle	LAB	Nepal, Darjeeling hills, Sikkim
Mesu	Bamboo shoot	Sour; pickle	LAB	Nepal, Darjeeling hills, Sikkim, Bhutan
Soibum	Bamboo shoot	Sour-acidic; curry	LAB	Manipur
Soidon	Bamboo shoot tips	Sour-acidic; curry	LAB	Manipur
Soijim	Bamboo shoot	Liquid, sour; condiment	LAB	Manipur
Ekung	Bamboo shoot	Sour-acidic; curry, soup	LAB	Arunachal Pradesh
Hirring	Bamboo shoot tips	Sour-acidic; curry, soup	LAB	Arunachal Pradesh
Eup	Bamboo shoot	Dry, acidic; curry, soup	LAB	Arunachal Pradesh
Lung-siej	Bamboo shoot	Sour-acidic; curry	LAB	Meghalaya
Bastanga	Bamboo shoot	Sour-acidic; curry	Unknown	Nagaland
Sinnamani	Radish	Freshly fermented, sour; pickle	Unknown	Nepal
Anishi	Taro leaves	Fermented; sour; curry	Unknown	Nagaland

2. Fermented legumes

Kinema	Soybean	Sticky, flavored; curry	*Bacillus subtilis*	East Nepal, Darjeeling hills, Sikkim, Bhutan
Hawaijar	Soybean	Sticky, flavored; side dish as fish substitute	*Bacillus* spp.	Manipur
Tungrymbai	Soybean	Sticky, flavored; curry	*Bacillus* spp.	Meghalaya
Aakhone	Soybean	Sticky or dry-cakes; side dish	*Bacillus* spp.	Nagaland
Bekang	Soybean	Sticky, flavored; side dish	*Bacillus* spp.	Mizoram
Peruyaan	Soybean	Sticky, soybeans; curry	*Bacillus* spp.	Arunachal Pradesh
Maseura	Black gram	Dry, ball-like; condiment	Bacilli, LAB and yeasts	Nepal, Darjeeling hills, Sikkim

3. Fermented milk products

Dahi	Cow milk	Curd; savory	LAB, yeasts	All
Shyow	Yak milk	Curd; savory	LAB, yeasts	Sikkim, Bhutan, Tibet, Ladakh
Gheu	Cow milk	Butter	LAB, yeasts	All
Maa	Yak milk	Butter	LAB, yeasts	Sikkim, Bhutan, Tibet, Ladakh
Mohi	Cow milk	Buttermilk; refreshing beverage	LAB, yeasts	Eastern Himalayas
Lassi	Cow milk	Buttermilk; refreshing beverage	LAB, yeasts	Western Himalayas
Chhurpi (soft)	Cow milk	Soft, cheese-like; curry, pickle	LAB, yeasts	Nepal, Darjeeling hills, Sikkim
Chhurpi (hard)	Yak milk	Hard-mass, masticator	LAB, yeasts	Nepal, Darjeeling hills, Sikkim, Bhutan, Tibet, Ladakh, Arunachal Pradesh, Himachal Pradesh

(continued on next page)

Table 1.1 (continued) Himalayan Fermented Foods and Beverages

Food	Substrate	Nature of product	Microbes	Region
Dudh chhurpi	Cow milk	Hard-mass, masticator	LAB, yeasts	Nepal, Darjeeling hills, Sikkim, Bhutan
Phrung	Yak milk	Hard-mass, masticator	Unknown	Arunachal Pradesh
Chhu or sheden	Cow/yak milk	Soft, strong flavored; curry	LAB, yeasts	Sikkim, Bhutan
Chur yuupa	Yak milk	Soft, flavored; curry, soup	Unknown	Arunachal Pradesh
Somar	Cow/yak milk	Paste, flavored; condiment	LAB	Nepal, Darjeeling hills, Sikkim
Dachi	Cow/yak milk	Soft, cheese-like, strong flavored; hot curry	Unknown	Bhutan
Philu	Cow/yak milk	Cream; fried curry with butter	LAB	Sikkim, Bhutan, Tibet
Pheuja or Suja	Tea, yak butter	Fermented butter tea	Unknown	Sikkim, Bhutan, Tibet, Ladakh
Paneer	Whey of cow milk	Soft, cheese-like product; fried snacks	LAB	Western Himalayas
4. Fermented cereals				
Selroti	Rice, wheat flour, milk	Pretzel-like, deep fried; bread	Yeasts, LAB	Nepal, Darjeeling hills, Sikkim, Bhutan
Jalebi	Wheat flour	Crispy sweet, deep fried pretzels; snacks	Yeasts, LAB	Western Himalayas
Nan	Wheat flour	Leavened bread; baked; staple	Yeasts, LAB	Western Himalayas
Sidu	Wheat flour, opium seeds, walnut	Steamed bread, oval-shaped; staple	Unknown	Himachal Pradesh, Uttarakhand

	Substrate	Usage	Microorganisms	Region
Chilra	Wheat, barley, buckwheat	Like dosa; staple	LAB, yeasts	Himachal Pradesh, Uttarakhand
Marchu	Wheat flour	Baked breads; staple	Unknown	Himachal Pradesh
Bhaturu	Wheat flour	Baked breads; staple	LAB, yeasts	Himachal Pradesh, Uttarakhand
Seera	Wheat grains	Dried, sweet dish	Unknown	Himachal Pradesh
5. Fermented/dry fish products				
Suka ko maacha	River fish	Smoked, sun-dried; curry	LAB, *Bacillus*, yeasts	Nepal, Darjeeling hills, Sikkim
Gnuchi	River fish	Smoked; curry	LAB, *Bacillus*, yeasts	Darjeeling hills, Sikkim
Sidra	Fish	Dried fish; curry	LAB, yeasts	Nepal, Darjeeling hills, Sikkim, Bhutan
Sukuti	Fish	Dried fish; curry	LAB, yeasts	Nepal, Darjeeling hills, Sikkim, Bhutan
Ngari	Fish	Fermented fish; curry	LAB, yeasts	Manipur
Hentak	Fish and petioles of aroid plants	Fermented fish paste; curry	LAB, yeasts	Manipur
Tungtap	Fish	Fermented; pickle	LAB, yeasts	Meghalaya
Karati	Fish	Dried, salted; curry	LAB, yeasts	Assam
Bordia	Fish	Dried, salted; curry	LAB, yeasts	Assam
Lashim	Fish	Dried, salted; curry	LAB, yeasts	Assam
Mio	Fish	Dried; curry	Unknown	Arunachal Pradesh
Naakangba	Fish	Sun-dried; pickle, curry	Unknown	Manipur
Ayaiba	Fish	Smoked fish; pickle, curry	Unknown	Manipur

(continued on next page)

Table 1.1 (continued) Himalayan Fermented Foods and Beverages

Food	Substrate	Nature of product	Microbes	Region
6. Fermented/dry meat products				
Lang kargyong	Beef	Sausage (soft or hard, brownish); curry	LAB	Sikkim, Darjeeling hills, Bhutan, Tibet, Arunachal Pradesh, Ladakh
Yak kargyong	Yak	Sausage (soft, brownish); curry	LAB	Sikkim, Bhutan, Tibet, Arunachal Pradesh, Ladakh
Faak kargyong	Pork	Sausage (soft or hard, brownish); curry	LAB	Sikkim, Darjeeling hills, Bhutan, Tibet
Lang satchu	Beef meat	Dried or smoked meat (hard, brownish); curry	LAB	Sikkim, Darjeeling hills, Bhutan, Tibet, Ladakh
Yak satchu	Yak meat	Dried or smoked meat (hard, brownish); curry	LAB	Sikkim, Darjeeling hills, Bhutan, Tibet, Arunachal Pradesh, Ladakh
Suka ko masu	Buffalo meat	Dried or smoked meat (hard, brownish chocolate); curry	LAB	Nepal, Darjeeling hills, Sikkim
Yak chilu	Yak fat	Hard, used as substitute of an edible oil	LAB	Sikkim, Bhutan, Tibet, Arunachal Pradesh, Ladakh
Lang chilu	Beef fat	Hard, used as an edible oil	LAB	Sikkim, Bhutan, Tibet, Ladakh
Luk chilu	Sheep fat	Hard, used as an edible oil	LAB	Sikkim, Bhutan, Tibet, Ladakh, Nepal
Yak kheuri	Yak	Chopped intestine of yak; curry	LAB	Sikkim, Bhutan, Tibet, Arunachal Pradesh, Ladakh
Lang kheuri	Beef	Chopped intestine of beef; curry	LAB	Sikkim, Bhutan, Tibet, Arunachal Pradesh, Ladakh
Chartayshya	Chevon	Dried/smoked meat; curry	LAB, bacilli, micrococci, yeasts	Kumaon hills of Uttarakhand

Jannua	Small intestine of chevon, finger millet	Sausage (soft); curry	LAB, bacilli, micrococci, yeasts	Kumaon hills, western Nepal
Arjia	Large intestine of chevon	Sausage; curry	LAB, bacilli, micrococci, yeasts	Kumaon hills, western Nepal
Bagjinam	Pork	Fermented pork; curry	Unknown	Nagaland
Sukula	Buffalo	Dried / smoked meat; curry	Unknown	Nepal
7. Mixed-starter cultures				
Marcha	Rice, wild herbs, spices	Dry, mixed starter to ferment alcoholic beverages	Molds, yeasts, LAB	Eastern Himalayas
Phab	Wheat, wild herbs	Dry, mixed starter to ferment alcoholic beverages	Molds, yeasts, LAB	Sikkim, Bhutan, Tibet, Ladakh
Ipoh/Siye	Rice, wild herbs	Starter to ferment alcoholic beverages	Unknown	Arunachal Pradesh
Hamei	Rice, wild herbs	Dry, mixed starter to ferment alcoholic beverages	Molds, yeasts, LAB	Manipur
Mana	Wheat, herbs	Starter to ferment alcoholic beverages	*Aspergillus oryzae*	Nepal
Manapu	Rice-wheat, herbs	Starter to ferment alcoholic beverages	Yeasts and molds	Nepal
Emao	Rice-herbs	Starter to ferment alcoholic beverages	Unknown	Assam
Thiat	Rice-herbs	Starter to ferment alcoholic beverages	Unknown	Meghalaya

(continued on next page)

Table 1.1 (continued) Himalayan Fermented Foods and Beverages

Food	Substrate	Nature of product	Microbes	Region
8. Alcoholic beverages and distilled liquor				
Pham	Rice–herbs	Starter to ferment alcoholic beverages	Unknown	Arunachal Pradesh
Khekhrii	Germinated rice	Starter to ferment *zutho/zhuchu*	Unknown	Nagaland
Balan	Wheat	Starter to ferment alcoholic beverages	Unknown	Uttarakhand
Bakhar	Rice–herbs	Starter to ferment alcoholic beverages	Yeasts	Western Himalayas
Kodo ko jaanr	Finger millet	Mildly alcoholic, slightly sweet-acidic; alcoholic beverage	Yeasts, LAB	Eastern Himalayas
Chyang/ Chee	Finger millet/ barley	Mildly alcoholic, slightly sweet-acidic; alcoholic beverage	Yeasts, LAB	Sikkim, Bhutan, Tibet, Ladakh
Bhaati jaanr	Rice	Mildly alcoholic, sweet-sour, food beverage	Yeasts, LAB	Nepal, Darjeeling hills, Sikkim
Makai ko jaanr	Maize	Mildly alcoholic, sweet-sour, food beverage	Yeasts, LAB	Nepal, Darjeeling hills, Sikkim
Gahoon ko jaanr	Wheat	Mildly alcoholic, slightly acidic; alcoholic beverage	Yeasts, LAB	Nepal, Darjeeling hills, Sikkim
Simal tarul ko jaanr	Cassava tuber	Mildly alcoholic, sweet-sour; food beverage	Yeasts, LAB	Nepal, Darjeeling hills, Sikkim
Jao ko jaanr	Barley	Mildly alcoholic, slightly acidic; alcoholic beverage	Yeasts, LAB	Nepal, Darjeeling hills, Sikkim
Faapar ko jaanr	Buckwheat	Mildly alcoholic, slightly acidic; alcoholic beverage	Yeasts, LAB	Nepal, Darjeeling hills, Sikkim

Atingba	Rice	Mildly alcoholic, sweet-sour, food beverage	Unknown	Manipur
Apong	Rice	Mildly alcoholic, beverage	Unknown	Arunachal Pradesh
Pona	Rice	Mildly alcoholic, sweet-sour, food beverage; paste	Unknown	Arunachal Pradesh
Ennog	Rice, paddy husk	Black rice beer	Unknown	Arunachal Pradesh
Jou	Rice	Alcoholic beverage	Unknown	Nagaland
Zutho/ Zhuchu	Rice	Milky white, alcoholic beverage	LAB, yeasts	Nagaland
Oh	Rice-millet	Soft, alcoholic beverage	Unknown	Arunachal Pradesh
Themsing	Finger millet/barley	Alcoholic beverage	Unknown	Arunachal Pradesh
Mingri	Maize-rice/barley	Alcoholic beverage	Unknown	Arunachal Pradesh
Lohpani	Maize-rice/barley	Alcoholic beverage	Unknown	Arunachal Pradesh
Zu	Rice	Alcoholic beverage	Unknown	Assam
Sura	Finger millet	Food beverage; staple	Unknown	Himachal Pradesh
Lugri	Barley	Alcoholic beverage	Yeasts	Himachal Pradesh, Ladakh, Tibet
Sing sing	Barley	Beverage	Yeasts	Ladakh
Buza	Barley	Thick liquor	…	Ladakh
Raksi	Cereals	Clear distilled liquor; alcoholic drink	Yeasts, LAB	Eastern Himalayas
Aara	Cereals	Clear distilled liquor; alcoholic drink	Unknown	Arunachal Pradesh
Duizou	Red rice	Alcoholic drink	Unknown	Nagaland
Nchiangne	Red rice	Distilled liquor	Unknown	Nagaland
Ruhi	Rice	Distilled liquor	Unknown	Nagaland
Madhu	Rice	Distilled liquor	Yeast, Mold	Nagaland
Yu	Rice	Distilled liquor	Unknown	Manipur

(continued on next page)

Table 1.1 (continued) Himalayan Fermented Foods and Beverages

Food	Substrate	Nature of product	Microbes	Region
Kiad-lieh	Rice	Distilled liquor	Unknown	Meghalaya
Poko	Rice	Food beverage	Yeasts, LAB	Nepal
Bhang-chyang	Maize-rice/barley	Extract of *mingri*; alcoholic beverages	Unknown	Arunachal Pradesh
Daru	Cereal	Alcoholic beverages; filtrate; jiggery	Unknown	Himachal Pradesh
Chulli	Apricot	Alcoholic beverages; filtrate; alcoholic drink	Unknown	Himachal Pradesh
9. Miscellaneous fermented products				
Achar or chutney	Fruits, vegetables, mixed with oil, salt	Acidic, hot and sour; pickles	LAB	The Himalayas
Black Tea	Tea	Nonalcoholic drink	Unknown	The Himalayas
Kombucha or tea fungus	Tea liquor	Nonalcoholic drink	LAB, Yeasts	Tibet, Ladakh
Chuk	Fruits	Dark-brown paste, sour bitter taste; therapeutic uses	Unknown	Nepal, Darjeeling hills, Sikkim
Hakua	Rice	Strong off-flavor; fermented paddy	Unknown	Nepal, Darjeeling hills, Sikkim
Crabs	Crabs	Side dish	Unknown	Nagaland

and consumed for centuries, probably for more than 2500 years. A few products are lesser known and confined to particular communities and specific regions in the Himalayas. The mountain women play an important role with their indigenous knowledge of food fermentation. Their participation spans from cultivation to harvesting, fermentation to culinary skills, and production to marketing. Rural women also sell the ethnic food products in local markets and earn their livelihood, thereby directly or indirectly enhancing the regional economy.

1.6.1 Microbiology of fermented foods

Fermented foods harness diverse functional microorganisms from the environment (Tamang 2002). Microorganisms are present in or on the ingredients, utensils, and the general environment, and they are selected through adaptation to the substrate for fermentation (Hesseltine 1983; Steinkraus 1997; Tamang 1998a). In the Indian subcontinent, mostly due to wide variation in agroclimatic conditions and the diverse forms of food culture of different ethnic peoples, a diversity of microorganisms is associated with traditional fermented foods and beverages (Soni and Sandhu 1990b; Tamang 1998a; Tamang and Holzapfel 1999). Microorganisms change the chemical composition of raw materials during food fermentation, enhancing the nutritive value; enriching the bland diet with improved flavor and texture; preserving the perishable foods; fortifying the products with essential amino acids, vitamins, and minerals; degrading the undesirable components and antinutritive factors; improving the digestibility; and stimulating the probiotic functions (Steinkraus 1994; Stiles and Holzapfel 1997; Adams and Nout 2001).

Any food researcher studying fermented food should accomplish the following objectives, such as determination of microbial profiles, isolation, enrichment, purification, phenotypic and genotypic characterization, proper identification of functional microorganisms, and preservation and deposition of identified strains in microbial culture collection centers. Authentic identification of functional microorganisms associated with the production of final edible products is an important aspect of microbial properties that determine the quality of the product (Tamang and Holzapfel 1999). Studies on the nutritional profile, technological or functional properties, food safety, process optimization, medicinal value, etc., are also important aspects of fermented foods. Genotypic identification using molecular methods such as DNA base composition, DNA hybridization, and ribosomal RNA and chemotaxonomical tools such as cell wall studies, cellular fatty acids, and isoprenoid quinones are helpful when the conventional approach of identification is not reliable. These ethnic fermented foods have been extensively studied, and the functional microorganisms have been isolated, properly characterized, and identified, based on modern phenotypic (API and Biolog systems) and genotypic characteristics (RAPD-PCR, rep-PCR, species-specific PCR techniques, 16S rRNA sequencing, DNA-DNA hybridization), and preserved in 15% glycerol at −20°C.

Three major groups of microorganisms are present or associated with ethnic fermented foods: bacteria (mostly lactic acid bacteria), yeasts, and mycelial fungi. Lactic acid bacteria are Gram-positive, nonsporing, catalase-negative, devoid of cytochromes, of nonaerobic habitat but aerotolerant, fastidious, acid-tolerant, and strictly fermentative, with lactic acid as the major end product during sugar fermentation (Axelsson 1998). LAB produce lactic acid during traditional fermentation, the characteristic fermentative product, which reduces the pH of the substrate to a level where growth of putrefactive, pathogenic, and toxinogenic bacteria are inhibited (Holzapfel et al. 1995). The LAB that are usually designated as GRAS (generally recognized as safe) status in foods (Donohue and Salminen 1996) can also exert a biopreservative effect (Holzapfel et al. 2003). The genera of LAB most often present in foods are *Carnobacterium, Enterococcus, Lactobacillus, Lactococcus, Leuconostoc, Oenococcus, Pediococcus, Streptococcus, Tetragenococcus, Vagococcus,* and *Weissella* (Stiles and Holzapfel 1997; Carr et al. 2002). The advantages of lactic acid food fermentations include resistance to spoilage and food toxins, which make the foods less likely to transfer pathogenic microorganisms. They also preserve the foods between the time of harvest and consumption, modify the flavor of the original ingredients, and improve the nutritional value (Nout and Ngoddy 1997; Salminen and Wright 1998). LAB are the dominant microorganisms in many traditional fermented foods. Among nonlactic bacteria, *Bacillus* is mostly encountered in many Asian fermented soybean foods and also in some African fermented foods. The most common species reported in fermented foods are *B. subtilis* and *B. natto,* along with *B. licheniformis, B. thuringiensis, B. megaterium,* etc. (Kiers et al. 2000). *Bacillus* is an endospore forming, rod-shaped, Gram-positive, catalase positive, motile, and aerobic to semi-anaerobic growing bacterium (Gordon 1973). Some *Bacillus* strains produce λ-polyglutamic acid (PGA), which is an amino acid polymer commonly present in fermented soybean foods, giving a sticky texture (Urushibata et al. 2002).

Yeasts play vital roles in the production of many traditional fermented foods and beverages across the world (Aidoo et al. 2006), signifying the food culture of the regions and the community (Tamang and Fleet 2009). About 21 major genera with several species of functional yeasts have been isolated from fermented foods and beverages across the world, and these include *Brettanomyces* (its perfect stage, *Dekkera*), *Candida, Cryptococcus, Debaryomyces, Galactomyces, Geotrichum, Hansenula, Hanseniaspora* (its asexual counterpart *Kloeckera*), *Hyphopichia, Kluyveromyces, Metschnikowia,*

Pichia, Rhodotorula, Saccharomyces, Saccharomycodes, Saccharomycopsis, Schizo-saccharomyces, Torulopsis, Trichosporon, Yarrowia, and *Zygosaccharomyces* (Kurtzman and Fell 1998; Pretorius 2000; Tsuyoshi et al. 2005; Romano et al. 2006; Tamang and Fleet 2009).

Fungi in fermented foods are relatively limited. Some of the common genera of mycelial fungi associated with fermented foods and beverages of the world are *Actinomucor, Mucor, Rhizopus, Amylomyces, Monascus, Neurospora, Aspergillus* and *Penicillium* (Hesseltine 1983, 1991; Samson 1993; Nout and Aidoo 2002). All species of *Aspergillus* and *Penicillium* are non-toxin-producing species in fermented foods (Hesseltine 1983).

Functional microorganisms present in fermented foods have many biological functions that enhance their health-promoting benefits. These biological functions include biopreservation of perishable foods, bioenrichment of nutritional value, enrichment of diet, protective properties, improvement in lactose metabolism, production of enzymes, antimicrobial properties, destruction of undesirable components, degradation of antinutritive factors, improved digestibility, production of antioxidants, anticarcinogenic property, therapeutic values, bioavailability of minerals, lowering of serum lipids (and cholesterol), immunological effects, prevention of infectious diseases, etc. (Tamang 2007a; Shah 2007; Liong 2008).

1.6.2   Consumption patterns of fermented foods and beverages

A food consumption survey is an indispensable tool for assessing the nutritional intake and pattern of food consumption by the community (Tee et al. 2004). Yonzan and Tamang (1998) for the first time conducted a brief survey on the consumption patterns of traditional fermented foods of Darjeeling hills and Sikkim in India. About 13% of fermented foods constitute the daily meal of the ethnic people of Sikkim and Darjeeling hills (Tamang et al. 2007b). The survey on food consumption patterns in Sikkim shows that 11.7% of rural people are vegetarians and 88.3% are nonvegetarians (Tamang et al. 2007b). Daily per capita consumption of ethnic fermented foods in Sikkim is shown in Table 1.2. The survey report shows that 67.7% of people prepare most of the ethnic fermented foods at home for consumption (Tamang et al. 2007b). Consumption of fermented foods to total foods is highest in the north district of Sikkim (18.6%), followed by the south (13.9%), west (13.7%), and east (7.5%) (Tamang et al. 2007b). Consumption of alcoholic beverages and meat products is higher among the Bhutia and Lepcha than the ethnic Nepali, who have heterogeneous food habits ranging from vegetarian Brahmin to nonvegetarian non-Brahmin castes.

Food consumption data are needed to develop appropriate food-based dietary guidelines for the region and to monitor changes in dietary

Table 1.2 Daily Per Capita Consumption of Fermented Foods in Sikkim

Ethnic fermented foods	Per capita consumption by different ethnic people in Sikkim (g/day)			
	Nepali	Bhutia	Lepcha	Total in Sikkim
Kinema	3.4	1.1	1.4	2.3
Maseura	0.5	0	0	0.5
Gundruk	1.7	0.9	0.6	1.4
Sinki	1.3	0.5	0.2	1.1
Mesu	0.9	0.3	0.5	0.6
Khalpi	7.3	0	0.2	6.9
Dahi (ml)	37.1	32.8	26.5	34.1
Mohi (ml)	89.7	33.8	58.9	74.2
Gheu	7.1	14.9	2.8	7.9
Chhurpi (soft)	3.4	8.1	4.0	4.5
Chhurpi (hard)	0.02	0.8	0.01	0.3
Dudh chhurpi	0.03	0.01	0	0.01
Chhu	0	5.0	7.0	5.6
Somar	0	0.1	8.3	2.8
Philu	0	0.9	0.1	0.9
Suka ko masu	6.7	30.3	9.8	16.3
Selroti	8.4	3.6	5.9	8.0
Kodo ko jaanr	71.7	124.7	115.6	101.7
Makai ko jaanr	18.9	86.4	41.7	37.8
Bhaati jaanr	14.9	8.9	8.9	10.4
Gahoon ko jaanr	30.8	56.8	14.6	26.4
Raksi (ml)	33.7	203.8	7.9	57.7

behavior and patterns. Himalayan ethnic fermented foods and beverages play an important role as sources of protein, calories, minerals, and vitamins in the local diet. Because such ethnic fermented foods are rooted in a long tradition of dietary culture, their consumption is unlikely to be changed over a short period of time.

chapter two

Fermented vegetables

Vegetables are eaten in the Himalayas as a side dish at every meal. Green and leafy vegetables are grown on terraces in the hilly slopes from subtropical to temperate regions in the Himalayas. Some of the common vegetables grown in the Himalayas are fresh leaves of *rayo-sag* (*Brassica rapa* subsp. *campestris* var. *cuneifolia*); leafy mustard; radish; cabbage; cauliflower; young tendrils, fruits, and tubers of squash (*iskus*); *brinjal*; chilli; cucumber; young tendrils and fruits of pumpkin; tomato; tree tomato; lemon; local varieties of chilli; potato; beetroots; sweet potato; cassava; arum/taro; yam; ginger; turmeric; large cardamom; carrot; etc. Bamboo is an important plant for many ethnic people of the subtropical regions of the Himalayas as foods and as nonfood products, such as house materials, handicrafts, etc. (Tamang 2001a; Bhatt et al. 2003, 2005). Some common species of domesticated and wild bamboo tender shoots used as foods of the Himalayas are *Dendrocalamus hamiltonii*, *Dendrocalamus sikkimensis*, *Dendrocalamus giganteus*, *Bambusa tulda*, *Melocanna bambusoides*, *Phyllostachys assamica*, etc. Perishable and seasonal leafy vegetables; radishes; cucumbers; and young, edible, tender bamboo shoots are traditionally fermented into edible products in the Himalayas by the ethnic people using their indigenous knowledge of biopreservation.

2.1 Important fermented vegetables

Some ethnic fermented vegetable products of the Himalayas are: *gundruk, sinki, goyang,* and *khalpi* of Nepal, Darjeeling hills, Sikkim, and Bhutan; and *inziangsang* and *anishi* of Nagaland. Fermented bamboo-shoot products are: *mesu* of Nepal, Darjeeling hills, Sikkim, and Bhutan; *soidon, soibum,* and *soijim* of Manipur; *ekung, eup,* and *hiring* of Arunachal Pradesh; and *lung-siej* of Meghalaya.

2.1.1 Gundruk

Gundruk is an ethnic fermented vegetable, a dry and acidic product indigenous to the Nepali living in the Himalayan regions of India, Nepal, and Bhutan. *Gundruk* is generally produced during December to February, when large quantities of leaves of mustard, *rayo-sag* (local variety of mustard), radish, and other vegetables pile up, much more than people could consume fresh (Dietz 1984; Tamang 1997). The word *gundruk* has been derived from the Newari (Newar/Pradhan is one of the major castes of the Nepali) dialect *gunnu*, meaning "dried taro (*Colocasia*) stalk." *Gundruk* is similar to fermented acidic vegetable products of other countries, such as *kimchi* of Korea (Cheigh and Park 1994), *sauerkraut* of Germany (Breidt et al. 1995), *sunki* of Japan (Itabashi 1986), *pak-gard-dong* of Thailand (Mingmuang 1974), and *suan-cai* of China (Miyamoto et al. 2005).

2.1.1.1   Indigenous knowledge of preparation

In Darjeeling hills, Sikkim, and southern Bhutan, *gundruk* is prepared from fresh leaves of a local vegetable called *rayo-sag* (*Brassica rapa* subsp. *campestris* var. *cuneifolia*), mustard (*Brassica juncea*), and cauliflower (*Brassica oleracea* var. *botrytis*); the leaves are wilted and shredded using a sickle or knife. Wilted and shredded leaves are crushed mildly and pressed into an earthen jar or container, and made airtight by covering with dried bamboo bract sheaths and fern fronds that are weighted down by stones. The container is kept in a warm place and allowed to ferment naturally for about 7–10 days. A mild acidic taste indicates the completion of the fermentation, and the *gundruk* is removed from the container. Unlike other Asian fermented vegetables, freshly fermented *gundruk* is sun dried for 3–4 days (Figures 2.1 and 2.2). Dried *gundruk* can be preserved for 2 years or more at room temperature for consumption (Figure 2.3) (Tamang 2000b).

The preparation of *gundruk* in Nepal is slightly different. Fresh leaves of mustard, radish, and cauliflower are wilted for 2–3 days (Dahal et al. 2005). The leaves are shredded, pressed into an earthen jar and covered with lukewarm (30°C–35°C) water, and fermented at 16°C–20°C for 5–7 days. After fermentation, leaves are removed from the jar and sun dried for 3 days (Karki 1986). *Gundruk* preparation is usually done in winter, when the weather is less humid.

For large production of *gundruk*, pit fermentation is practiced in the Himalayan villages. A 2–3-ft-deep pit of same diameter is dug in a dry place, and the pit is cleaned, plastered with mud, and warmed by burning. After removal of the ashes, the pit is lined with bamboo sheaths and paddy straw. Crushed withered leaves are dipped in lukewarm water, squeezed and pressed tightly into the pit, and then covered with dry leaves that are weighted down by stones or heavy planks. The top of the pit is plastered with mud and left to ferment naturally for 15–22 days. Freshly fermented *gundruk* is removed, sun dried for 3–5 days, and stored at room temperature for future consumption.

2.1.1.2   Culinary practices and economy

Daily per capita consumption of *gundruk* is 1.4 g, with annual production of 3.2 kg per household in Sikkim (Tamang et al. 2007b). *Gundruk* is eaten as a soup or pickle. Soup is made by soaking *gundruk* in water

Leafy vegetable

↓

Let wilt (1–2 d)

↓

Crush mildly, soak briefly in warm water

↓

Pack into a pot, press tightly

↓

Ferment naturally for 15–22 d

↓

Take out freshly fermented *gundruk* and sun dry for 2–4 d

↓

GUNDRUK (dry)

Figure 2.1 Indigenous method of *gundruk* preparation in Darjeeling hills and Sikkim.

Figure 2.2 Nepali lady is sun drying fresh *gundruk* in Sikkim.

Figure 2.3 Dried *gundruk*.

for 10 min, squeezing out excess moisture, and frying in edible oil with chopped onions, tomatoes, chillies, turmeric powder, and salt. It is then boiled for 10–15 min, and served hot with steamed rice. *Gundruk* soup is a good appetizer in a bland and starchy diet.

   Gundruk is sold in all local markets. One kg of *gundruk* costs about Rs. 100 or more in markets. Rural women sell the product in local markets.

2.1.1.3   Microorganisms
Lactic acid bacteria (LAB): *Lactobacillus fermentum*, *Lb. plantarum*, *Lb. casei*, *Lb. casei* subsp. *pseudoplantarum*, and *Pediococcus pentosaceus* (Karki et al. 1983d; Tamang et al. 2005).

2.1.2   Sinki

Sinki, a fermented radish taproot product, is prepared by pit fermentation in the Himalayas. It is usually prepared during winter, when radish taproots are in plentiful supply. When the leaves of radish are fermented, it is called *gundruk*, and when the taproot is fermented, it is called *sinki*.

2.1.2.1   Indigenous knowledge of preparation
During *sinki* production, a 2–3-ft-deep pit of the same diameter is dug in a dry place. The pit is cleaned, plastered with mud, and warmed by burning. After removal of the ashes, the pit is lined with bamboo sheaths and

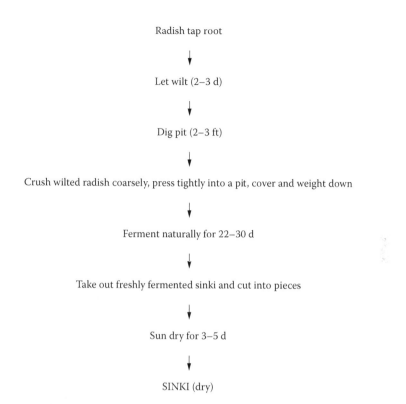

Radish tap root

↓

Let wilt (2–3 d)

↓

Dig pit (2–3 ft)

↓

Crush wilted radish coarsely, press tightly into a pit, cover and weight down

↓

Ferment naturally for 22–30 d

↓

Take out freshly fermented sinki and cut into pieces

↓

Sun dry for 3–5 d

↓

SINKI (dry)

Figure 2.4 Indigenous method of *sinki* preparation in Nepal.

paddy straw. Radish (*Raphanus sativus* L.) taproots are wilted for 2–3 days, crushed, dipped in lukewarm water, squeezed and pressed tightly into the pit, and then covered with dry leaves that are weighted down by heavy planks or stones. The top of the pit is plastered with mud and left to ferment naturally for 22–30 days (Figure 2.4). After completion of fermentation, fresh *sinki* is removed, cut into small pieces, sun dried for 3–5 days (Figure 2.5), and stored at room temperature for future consumption. Dry *sinki* can be kept for 2 years or more at room temperature (Figure 2.6). It is stored in an airtight container to avoid loss of typical *sinki* smell and to avoid absorption of moisture.

Pit fermentation of *sinki* preparation is a unique type of biopreservation of perishable radish by lactic acid fermentation in the Himalayas. Pit fermentation has been practiced in the South Pacific and Ethiopia for preservation of breadfruit, taro, banana, and cassava (Steinkraus 1996).

2.1.2.2   Culinary practices and economy
Sinki, with a highly acidic flavor, is typically used as a base for soup and as a pickle. The soup is made by soaking *sinki* in water for about 10 min,

Figure 2.5 Fresh *sinki* being sun dried in Sikkim.

Figure 2.6 Dried *sinki*.

squeezed to drain out excess water, and frying along with chopped onions, tomatoes, green chillies, and salt. Soup is served hot along with meals. Sinki soup is sourer in taste than that of gundruk. It is said to be a good appetizer, and people use it as a remedy for indigestion. The pickle is prepared by soaking *sinki* in water, squeezing it dry, and mixing it with salt, mustard oil, and chillies. Sinki pickle is more preferred by the Nepali than soup or curry.

Sinki is also sold in all local markets, mostly by the rural women for their livelihood. A kilogram of *sinki* costs about Rs. 150 in the local markets in Sikkim, Darjeeling hills, and southern Bhutan.

2.1.2.3 Microorganisms

Lactic acid bacteria (LAB): *Lactobacillus plantarum, Lb. brevis, Lb. casei*, and *Leuconostoc fallax* (Tamang and Srakar 1993; Tamang et al. 2005).

2.1.3 Goyang

Goyang is an ethnic fermented, slightly acidic vegetable food of the Sherpa of Sikkim and Nepal (Figure 2.7). It is prepared during rainy season, when the leaves of the wild plant, locally called *magane-saag* (*Cardamine*

Figure 2.7 *Goyang.*

macrophylla Willd.), belonging to the family Brassicaceae, are in plentiful supply in the hills of Nepal and Sikkim.

2.1.3.1   Indigenous knowledge of preparation

Leaves of a wild edible plant, locally called *magane-saag*, are collected, washed, and cut into pieces, then squeezed to drain off excess water, and are tightly pressed into bamboo baskets lined with two to three layers of leaves of fig plants. The tops of the baskets are then covered with fig plant leaves and fermented naturally at room temperature (15°C–25°C) for 25–30 days (Figure 2.8). Nowadays, poly bags or glass jars are used instead of bamboo baskets, as they are better at maintaining anaerobic condition during preparation of *goyang*. After completion of desired fermentation, fresh *goyang* is transferred into an airtight container, where it can be stored for 2–3 months. However, the shelf life of *goyang* can be prolonged by making the freshly fermented *goyang* into balls and sun drying for 2–3 days. Sun-dried *goyang* can be kept for several months at room temperature.

2.1.3.2   Culinary practices and economy

Goyang is generally prepared at home by the Sherpa women; there is no report of selling in the local markets. *Goyang* is boiled in a soup along with

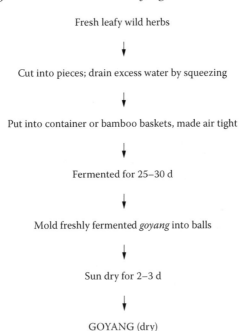

Fresh leafy wild herbs

↓

Cut into pieces; drain excess water by squeezing

↓

Put into container or bamboo baskets, made air tight

↓

Fermented for 25–30 d

↓

Mold freshly fermented *goyang* into balls

↓

Sun dry for 2–3 d

↓

GOYANG (dry)

Figure 2.8 Indigenous method of *goyang* preparation in Sikkim.

yak or beef meat and noodles to make a thick *thukpa*, a common staple food of the Sherpa (Tamang 2006).

2.1.3.3   Microorganisms
Lactic acid bacteria (LAB): *Lactobacillus plantarum*, *Lb. brevis*, *Lactococcus lactis*, *Enterococcus faecium*, and *Pediococcus pentosaceus*; yeasts *Candida* spp. (Tamang and Tamang 2007).

2.1.4   Inziangsang

Inziangsang or *ziangsang* (Figure 2.9) is a traditional fermented leafy vegetable product of Nagaland and Manipur that is mostly consumed by the Naga. It is prepared from mustard leaves and is very similar to *gundruk*.

2.1.4.1   Indigenous knowledge of preparation
Withered mustard leaves, locally called *hangam* (*Brassica juncea* [L.] Czern), are collected, crushed, and soaked in warm water. Leaves are squeezed to remove excess water and pressed into the container and made airtight to maintain the anaerobic condition. The container is kept at ambient temperature (20°C–30°C) and allowed to ferment for 7–10 days (Figure 2.10). Like *gundruk*, freshly prepared *inziangsang* is sun dried for 4–5 days and stored in a closed container for a year or more at room temperature for future consumption. Freshly fermented *inziangsang* juice is also extracted, instead of sun drying, by squeezing with hand, and the juice

Figure 2.9 *Inziangsang.*

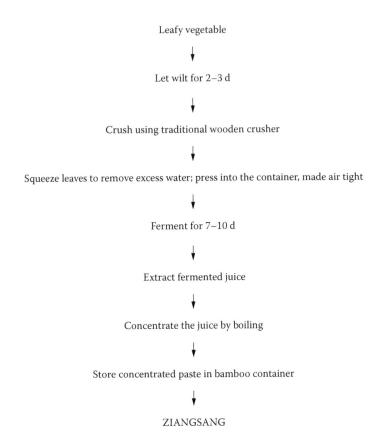

Leafy vegetable

↓

Let wilt for 2–3 d

↓

Crush using traditional wooden crusher

↓

Squeeze leaves to remove excess water; press into the container, made air tight

↓

Ferment for 7–10 d

↓

Extract fermented juice

↓

Concentrate the juice by boiling

↓

Store concentrated paste in bamboo container

↓

ZIANGSANG

Figure 2.10 Indigenous method of *inziangsang* preparation in Manipur.

is concentrated by boiling. The liquid form of fermented extract is called *ziang dui*, and the concentrated paste is called *ziangsang*. Extract concentrate *ziangsang* is stored in traditional bamboo containers for a year.

2.1.4.2   Culinary practices and economy
Inziangsang is consumed as a soup by the Naga with steamed rice. Fermented extract *ziang dui* is used as a condiment in local meals. *Inziangsang* is sold in the local market in Manipur and Nagaland.

2.1.4.3   Microorganisms
Lactic acid bacteria (LAB): *Lactobacillus plantarum, Lb. brevis,* and *Pediococcus acidilactici* (Tamang et al. 2005).

2.1.5 *Khalpi*

Khalpi or *khaipi* is a nonsalted fermented cucumber (*Cucumis sativus* L.) product, consumed by the Brahmin Nepali in Nepal, Darjeeling hills, and Sikkim. It is the only reported fermented cucumber product in the entire Himalayan region.

2.1.5.1 *Indigenous knowledge of preparation*

Ripened cucumber is collected from the field and cut into suitable pieces. The pieces of cucumber are sun dried for 2 days and then put into a bamboo vessel (Figure 2.11), locally called *dhungroo*, and made airtight by covering with dried leaves. It is fermented naturally at room temperature for 3–5 days (Figure 2.12). Fermentation after 5 days makes the product sourer in taste, which is not preferred by the consumers. *Khalpi* is prepared in the months of September and October.

2.1.5.2 *Culinary practices*

Khalpi is consumed as a pickle by adding mustard oil, salt, and powdered chillies in meal with steamed rice. It is not sold in markets.

Figure 2.11 *Khalpi* preparation by Nepali woman.

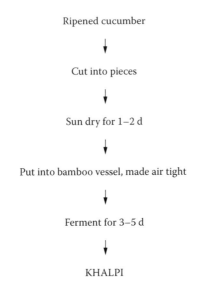

Ripened cucumber

↓

Cut into pieces

↓

Sun dry for 1–2 d

↓

Put into bamboo vessel, made air tight

↓

Ferment for 3–5 d

↓

KHALPI

Figure 2.12 Indigenous method of *khalpi* preparation in Sikkim.

2.1.5.3   Microorganisms

Lactic acid bacteria (LAB): *Lactobacillus plantarum*, *Lb. brevis*, and *Leuconostoc fallax* (Tamang et al. 2005).

2.1.6   Mesu

Mesu is a traditional fermented bamboo-shoot pickle with sour-acidic taste. The word *mesu* is directly derived from the Limboo language, in which *me* means "young bamboo shoot" and *su* means "sour" (Tamang 1998b). The Lepcha call it *satit*. The unfermented bamboo shoot in Nepali is known as *tama*, which is a very popular vegetable curry in Nepal, Darjeeling hills, Sikkim, and southern Bhutan.

2.1.6.1   Indigenous knowledge of preparation

Young edible shoots of some common bamboo species—locally called *choya bans* in the Nepali language (*Dendrocalamus hamiltonii*), *karati bans* (*Bambusa tulda*), and *bhalu bans* (*Dendrocalamus sikkimensis*)—are defoliated, chopped finely, and pressed tightly into a green hollow bamboo stem. (Figure 2.13) The tip of the vessel is covered tightly with leaves of bamboo or other wild plants and left to ferment under natural anaerobic conditions for 7–15 days (Figure 2.14). Completion of fermentation is indicated by the typical *mesu* flavor and taste.

Figure 2.13 *Mesu* in green bamboo vessel with cap made of bamboo leaves.

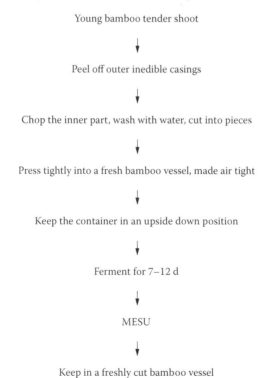

Young bamboo tender shoot

↓

Peel off outer inedible casings

↓

Chop the inner part, wash with water, cut into pieces

↓

Press tightly into a fresh bamboo vessel, made air tight

↓

Keep the container in an upside down position

↓

Ferment for 7–12 d

↓

MESU

↓

Keep in a freshly cut bamboo vessel

Figure 2.14 Flow sheet of *mesu* preparation in Darjeeling hills.

2.1.6.2   Culinary practices and economy

Mesu is eaten as a pickle. *Mesu* pickle is mixed with oil, chillies, and salt, and can be kept in a closed jar for several months without refrigeration. It is sold during the rainy season in the local markets when bamboo shoots are plentifully available. *Mesu* is sold in green bamboo vessels; each costs about Rs. 10–15.

2.1.6.3   Microorganisms

Lactic acid bacteria (LAB): *Lactobacillus plantarum, Lb. brevis, Lb. curvatus, Leuconostoc citreum*, and *Pediococcus pentosaceus* (Tamang and Sarkar 1996; Tamang et al. 2008).

2.1.7   Soibum

Soibum is an ethnic fermented tender bamboo-shoot food of Manipur and neighboring states in North East India. *Soibum* is exclusively produced from succulent bamboo shoots belonging to *Dendrocalamus hamiltonii, D. sikkimensis, D. giganteus, Melacona bambusoides, Bambusa tulda*, and *B. balcona*.

2.1.7.1   Indigenous knowledge of preparation

In Manipur, there are two types of traditional methods of preparation of *soibum*: (1) Noney/Kwatha type and (2) Andro type.

1. The Noney/Kwatha type of *soibum* production is found to be more common and popular. In this type, batch fermentation with more acidic taste is carried out in traditionally designed bamboo chambers. Nowadays, the traditional lining of bamboo chambers with forest leaves is being replaced with polyethylene sheets (Figure 2.15). The thin slices of the succulent bamboo shoots are packed compactly into this chamber. After filling the chamber with the slices to its capacity, the upper surface is sealed with a polyethylene sheet, and weights are then put on top for proper pressing. Production of good-quality *soibum* can be achieved with adequate pressing. The bottom of the chamber is perforated for draining acidic fermented juice during fermentation. It is left for 6–12 months for fermentation. After fermentation, *soibum* is stored for 10–12 months.
2. The Andro type of preparation of *soibum* is practiced only in Andro village in Manipur in the bulky roasted earthen pot by fed-batch fermentation. In this method, the people fill a portion of the pot with the bamboo-shoot slices and allow for fermentation. When the fermentation occurs, the mash volume is reduced, and additional quantity of fresh bamboo slices are added. After every addition of the slices, slight pressure is given with the hand. This process is repeated until the pot

Young bamboo shoot

↓

Defoliate, chop slices; wash with water

↓

Press tightly into the bamboo or earthen jar

↓

Ferment for 3–12 months

↓

SOIBUM

Figure 2.15 Indigenous method of *soibum* preparation in Manipur.

is filled with the bamboo-shoot slices and fermented for 6–12 months. In this method, the fermented juice is not allowed to drain out.

In all types of *soibum*, the aging of the preparation is very important. When the incubation is prolonged, the quality of the *soibum* is improved, and the traditional consumers prefer the flavor.

2.1.7.2   Culinary practices and economy

A variety of recipes are prepared by the Meitei from *soibum*, such as

Ironba: boiled *soibum* along with potato, taro petiole, *Eurale ferox* seeds, and coriander as condiments, and as a pickle made with chillies and *ngari* (fermented fish)

Athongba: cooked *soibum* with vegetables, particularly pumpkin, with basil leaves as a curry, or cooked *soibum* with fish/egg as a curry, or cooked *soibum* with pork as a favorite dish of the Kuki and Naga and eaten with cooked rice

Kangou: fried *soibum* after washing with *chenghi* (rice-wash water) with fish, potato, and vegetables as a side dish

Chagempomba: cooked *soibum* with *mimosa*, *parkia* beans, young leaves of pea, and *hawaijar* (fermented soybean paste) as a side dish. This *chagempomba* recipe is occasionally consumed as a special item by the Meitei in Manipur.

Soibum is commonly sold in local vegetable markets in Manipur (Figure 2.16). It costs about Rs. 40–50 per kilogram. *Soibum* production is an unorganized sector in the food industry in Manipur. There are a few producers who make an approximate 121% profit by selling *soibum* to

Figure 2.16 *Soibum* is being sold in Imphal markets in Manipur.

marginal buyers. The estimated annual income from the sale of *soibum* is approximately Rs. 230,000 (Tamang, Jeyaram, and Rangita, unpublished).

2.1.7.3   Microorganisms

Lactic acid bacteria (LAB): *Lactobacillus plantarum, Lb. brevis, Lb. coryniformis, Lb. delbrueckii, Leuconostoc fallax, Leuc. lactis, Leuc. mesenteroides, Enterococcus durans, Streptococcus lactis*; bacilli: *Bacillus subtilis, B. licheniformis, B. coagulans,* and *Micrococcus luteus*; yeasts: *Candida, Saccharomyces,* and *Torulopsis* spp. (Giri and Janmey 1987; Sarangthem and Singh 2003; Tamang et al. 2008).

2.1.8   Soidon

Soidon is a fermented product of Manipur made from the tip of mature bamboo shoots. It is a very popular fermented bamboo-shoot product in the diet of the Meitei in Manipur.

2.1.8.1   Indigenous knowledge of preparation

The tips or apical meristems of mature bamboo shoots (*Bambusa tulda, Dendrocalamus giganteus,* and *Melocana bambusoides*) are collected, and the outer casings and lower portions are removed. Whole tips are submerged

Figure 2.17 *Soidon* of Manipur.

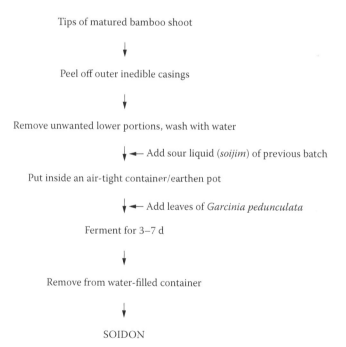

Figure 2.18 Indigenous method of *soidon* preparation in Manipur.

in water in an earthen pot (Figure 2.17). The sour liquid (*soijim*) of a previous batch is added as starter in 1:1 dilution, and the preparation is covered and fermented for 3–7 days at room temperature (Figure 2.18). Leaves of *Garcinia pedunculata* Roxb. (family: Guttiferae), locally called *heibung* in the Manipuri language, may be added in the fermenting vessel during fermentation to enhance the flavor of *soidon*. After 3–7 days, *soidon* is removed from the pot and stored in a closed container at room temperature for a year.

2.1.8.2   Culinary practices and economy
Soidon is consumed as a curry as well as a pickle with steamed rice. It is sold in the local market in Manipur. It costs about Rs. 50–60 per kilogram.

2.1.8.3   Microorganisms
Lactic acid bacteria (LAB): *Lactobacillus brevis, Leuconostoc fallax,* and *Leuc. lactis* (Tamang et al. 2008).

2.1.9   Soijim

Soijim or *soijin* is the liquid formed during fermentation of *soidon*. It is acidic and sour in taste and is commonly used as a condiment in Manipur.

2.1.9.1   Indigenous knowledge of preparation
During fermentation of *soidon*, pieces of bamboo-shoot tips are merged in water in an earthen pot, covered, and fermented for 3–7 days at room temperature. After fermentation, pieces of bamboo-shoot tips called *soidon* are taken out, and the acidic liquid portion, called *soijim*, is kept in a bottle. *Soijim* can be kept in a closed container at room temperature for a year or more.

2.1.9.2   Culinary practices
Soijim is used as a starter for fermentation of *soidon* and as a condiment to supplement the sour taste in curry in Meitei cuisine.

2.1.9.3   Microorganisms
Lactic acid bacteria (LAB): *Lactobacillus brevis, Leuconostoc fallax,* and *Leuc. lactis* (Tamang et al. 2008).

2.1.10   Ekung

Ekung is an ethnic fermented bamboo tender-shoot product of Arunachal Pradesh. The word *ekung* is derived from the Nyishing dialect; the Adi call it *iku,* and the Apatani call it *hikku* in Arunachal Pradesh (Basar and Bisht

2002). It is prepared during mid April to early September, when the young bamboo shoots are in plentiful supply.

2.1.10.1 Indigenous knowledge of preparation

The ethnic people of Arunachal Pradesh collect locally grown, young, bamboo tender shoots (*Dendrocalamus hamiltonii, Bambusa balcooa, Dendrocalamus giganteus, Phyllostachys assamica, Bambusa tulda*), and the outer leaf sheaths are removed (Figure 2.19). The edible portions are chopped into very small pieces. A 3–4-ft-deep pit is dug in the forest, usually near a water source to facilitate washing of bamboo-shoot pieces. The bamboo baskets are laid into the pit and lined with leaves. Chopped bamboo-shoot pieces are put into the baskets. When the baskets are full, they are covered with leaves and then sealed. Heavy stones are kept to give weight to drain excess water from the bamboo shoots, which are fermented for 1–3 months (Figure 2.20). *Ekung* can be kept for a year in an airtight container at room temperature.

2.1.10.2 Culinary practices and economy

Ekung is consumed raw or is cooked with meat, fish, and vegetables by the Nyishing. It is sold in all the local markets of Arunachal Pradesh. Depending on the season, a kilogram of *ekung* costs Rs. 30–50.

Figure 2.19 *Ekung* of Arunachal Pradesh.

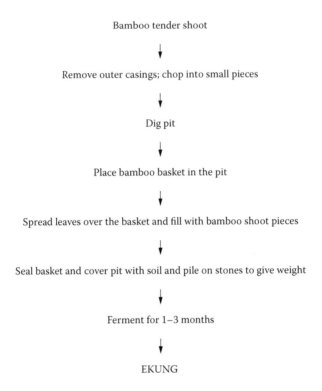

Bamboo tender shoot

↓

Remove outer casings; chop into small pieces

↓

Dig pit

↓

Place bamboo basket in the pit

↓

Spread leaves over the basket and fill with bamboo shoot pieces

↓

Seal basket and cover pit with soil and pile on stones to give weight

↓

Ferment for 1–3 months

↓

EKUNG

Figure 2.20 Indigenous method of *ekung* preparation in Arunachal Pradesh.

2.1.10.3   Microorganisms

Lactic acid bacteria (LAB): *Lactobacillus plantarum, Lb. brevis, Lb. casei,* and *Tetragenococcus halophilus* (Tamang and Tamang 2009).

2.1.11   Eup

Eup (Figure 2.21) is a dry-fermented bamboo tender-shoot food commonly prepared and consumed by different tribes of Arunachal Pradesh. *Eup,* derived from the Nyishing dialect, is a dry product and is known by different names in Arunachal Pradesh, such as *hi* by the Apatani, *nogom,* by the Khampti, and *ipe* by the Adi (Singh 2002).

2.1.11.1   Indigenous knowledge of preparation

Varieties of edible bamboo shoots are collected, and the outer casings are peeled off and washed. Bamboo shoots are chopped into small pieces and fermented in a similar manner as in *ekung*. Fermentation is completed in about 1–3 months (Figure 2.22). Unlike *ekung, eup* is a dry

Figure 2.21 *Eup* of Arunachal Pradesh.

product. After fermentation, the fermented product, now *eup*, is again cut into smaller pieces and then dried in the sun for 5–10 days until its color changes from whitish to chocolate brown. Usually *eup* is kept above the fireplace to avoid fungal attack, and can be consumed within 2 years.

2.1.11.2   Culinary practices and economy

Eup is consumed as a side dish with steamed rice, meat, fish, or vegetables. Curry of *eup* with meat is considered highly delicious by the people of Arunachal Pradesh. One kilogram of *eup* costs about Rs. 80–100. *Eup* is sold in the markets in Arunachal Pradesh throughout the year.

2.1.11.3   Microorganisms

Lactic acid bacteria (LAB): *Lactobacillus plantarum* and *Lb. fermentum* (Tamang and Tamang 2009).

2.1.12   Hirring

Hirring is a traditional fermented topmost whole bamboo-shoot product, commonly prepared by the Apatani tribes of Arunachal Pradesh. Nyishing tribes call it *hitch* or *hitak*. It is usually prepared along with *ekung*. It is prepared during June to early September.

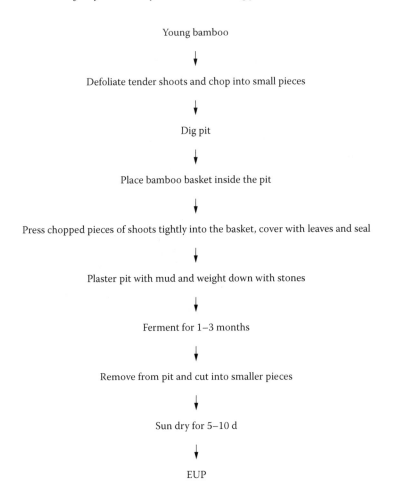

Young bamboo

↓

Defoliate tender shoots and chop into small pieces

↓

Dig pit

↓

Place bamboo basket inside the pit

↓

Press chopped pieces of shoots tightly into the basket, cover with leaves and seal

↓

Plaster pit with mud and weight down with stones

↓

Ferment for 1–3 months

↓

Remove from pit and cut into smaller pieces

↓

Sun dry for 5–10 d

↓

EUP

Figure 2.22 Indigenous method of *eup* preparation in Arunachal Pradesh.

2.1.12.1   Indigenous knowledge of preparation

Young tender bamboo shoots of *Dendrocalamus giganteus* or *Phyllostachys assamica* or *Bambusa tulda* are collected, and the outer leaf sheaths are removed. The topmost tender edible portions of the shoot are either cut longitudinally into two to three pieces of size 0.5–2 in. × 9–15 in., or whole shoots are flattened by crushing, and are put into bamboo baskets lined with leaves. The baskets are placed into the pit, covered with leaves, sealed and weighted down with heavy stones, and fermented for 1–3 months. After 1–3 months, the baskets are removed from the pits. *Hirring* that is ready for consumption is sour in taste and has an acidic flavor (Figure 2.23). *Hirring* can be kept for 2–3 months at room temperature.

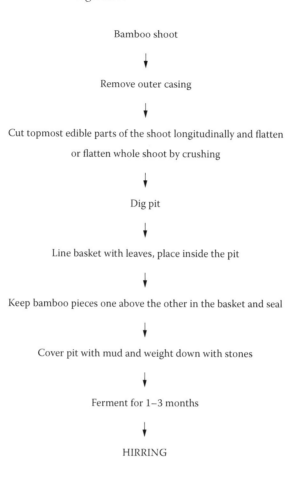

Bamboo shoot

↓

Remove outer casing

↓

Cut topmost edible parts of the shoot longitudinally and flatten
or flatten whole shoot by crushing

↓

Dig pit

↓

Line basket with leaves, place inside the pit

↓

Keep bamboo pieces one above the other in the basket and seal

↓

Cover pit with mud and weight down with stones

↓

Ferment for 1–3 months

↓

HIRRING

Figure 2.23 Indigenous method of preparation of *hirring* in Arunachal Pradesh.

2.1.12.2   *Culinary practices and economy*

Hirring is consumed as a side dish mixed with vegetables, meat, and fish, along with steamed rice. It is sold in the local markets (Figure 2.24) and costs about Rs. 150–200, depending on the season of availability of bamboo shoots.

2.1.12.3   *Microorganisms*

Lactic acid bacteria (LAB): *Lactobacillus plantarum* and *Lactococcus lactis* (Tamang and Tamang 2009).

Figure 2.24 *Hirring* of Arunachal Pradesh.

2.1.13   *Lung-siej*

Lung-siej is an ethnic fermented bamboo-shoot food of the Khasi in Meghalaya. It is prepared from *Dendrocalamus hamiltonii* species of bamboo grown in Meghalaya hills.

2.1.13.1   *Indigenous knowledge of preparation*

During production of *lung-siej*, tender bamboo shoots are selected. The bract sheaths are removed, washed thoroughly with water, and cleaned, and shoots are sliced into small pieces and pressed into either bamboo cylinders or inside a glass bottle (Agrahar-Murugkar and Subbulakshmi 2006). Bamboo cylinders are made by cutting the bamboo nodes in such a way that one side is closed while the other side remains open. Pieces of sliced bamboo shoots are filled inside the bamboo cylinder, which is closed with the help of leaves and sealed by tying up the rim of the cylinder with threads or grasses. The ends are sealed to prevent any accidental seepage of water into the cylinder, which would turn the shoots black, making the final product unfit for consumption. The bamboo cylinders are then immersed in the nearby stream upside down for a period of about 1–2 months for fermentation (Figure 2.25).

The use of glass bottles as a fermenting container is also common nowadays (Figure 2.26). In this process, the sliced shoots are pressed

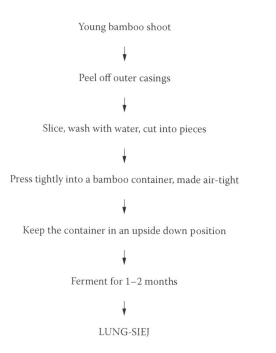

Young bamboo shoot

↓

Peel off outer casings

↓

Slice, wash with water, cut into pieces

↓

Press tightly into a bamboo container, made air-tight

↓

Keep the container in an upside down position

↓

Ferment for 1–2 months

↓

LUNG-SIEJ

Figure 2.25 Indigenous method of *lung-siej* preparation in Meghalaya.

Figure 2.26 *Lung-siej* of Meghalaya.

inside the glass bottle, and then water is added till they are submerged. Then the bottle is closed tightly with the cap and kept above the kitchen oven for 1 month. The keeping quality of bamboo shoots fermented inside the bottle is better than for the shoots fermented inside the bamboo cylinders. The shelf life of *lung-siej* prepared in a glass bottle is 10–12 months, whereas that for *lung-siej* prepared in bamboo cylinders is 1–2 months. Fermentation of the bamboo shoots in glass bottles is more popular among the people of urban areas, whereas fermentation of bamboo shoots inside the bamboo cylinder is preferred by rural people (Agrahar-Murugkar and Subbulakshmi 2006).

2.1.13.2   Culinary practices and economy

Lung-siej is eaten as a curry mixed with meats and fish along with steamed rice. It is sold in the local market in Meghalaya, and many Khasi women earn their livelihood by selling this product in Meghalaya.

2.1.13.3   Microorganisms

Lactic acid bacteria (LAB). These include species of *Lactobacillus* and *Pediococcus* (Tamang, unpublished).

2.1.14   Anishi

Anishi is an ethnic fermented vegetable product prepared from leaves of arum or taro (*Colocasia* spp.) and is used as a condiment by the Ao tribes of Nagaland.

2.1.14.1   Indigenous knowledge of preparation

Fresh mature green leaves of edible taro (*Colocasia* spp.) are collected and washed, and then the leaves are staked one above the other and wrapped finally with banana leaf. It is kept aside for about 6–7 days until the leaves turn yellow. The yellow leaves are mixed with chilli, salt, and ginger, and then ground into paste. The pastes are then molded into cakelike shapes that are then dried over the earthen oven in the kitchen for 2–3 days (Figure 2.27). The dried cakes are called *anishi* (Mao and Odyuo 2007).

2.1.14.2   Culinary practices

Anishi is used as a condiment and is cooked with dry meat, especially with pork, which is the favorite dish of the Ago tribe.

2.1.14.3   Microorganisms

Unknown.

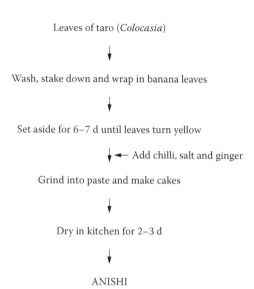

Leaves of taro (*Colocasia*)

↓

Wash, stake down and wrap in banana leaves

↓

Set aside for 6–7 d until leaves turn yellow

↓ ←— Add chilli, salt and ginger

Grind into paste and make cakes

↓

Dry in kitchen for 2–3 d

↓

ANISHI

Figure 2.27 Indigenous method of *anishi* preparation in Nagaland.

2.2 Microbiology

Lactic acid bacteria form a clear halo in $CaCO_3$-supplemented MRS (Man-Rogosa-Sharpe) agar (de Man et al. 1960) plates (Tamang et al. 2008). The lactic acid bacterial population in *gundruk, sinki, inziangsang, khalpi, sinki, mesu, soibum, soijim, ekung, eup,* and *hirring* is found in the viable numbers above 10^7 cfu/g (Table 2.1) comprising *Lactobacillus, Leuconostoc, Lactococcus, Pediococcus, Enterococcus,* and *Tetragenococcus* spp. (Tamang 2006).

Bacterial isolates are Gram-stained and tested for catalase production by placing a drop of 10% H_2O_2 solution on isolates, which are preliminarily grouped and identified on the basis of CO_2 production from glucose (Kandler 1983), ammonia production from arginine, growth at different temperatures, and the ability to grow in different concentrations of NaCl and levels of pH (Mundt 1986; Schillinger and Lücke 1987; Dykes et al. 1994; Simpson and Taguchi 1995; Wood and Holzapfel 1995). The configuration of lactic acid produced from glucose is determined enzymatically using D-lactate and L-lactate dehydrogenase test kits (Tamang et al. 2005). The presence of *meso*-diaminopimelic acid (DAP) in the cell walls of LAB is determined on cellulose plates using a thin-layer chromatography (Tamang et al. 2000). Sugar fermentation of LAB isolates is determined by the use of commercial API 50 CHL test strips (bioMérieux, France),

Table 2.1 Microbiological Populations of the
Himalayan Fermented Vegetable Products

Product	Log cfu/g sample		
	LAB[a]	Yeast[b]	AMC[c]
Goyang	6.8 ± 0.1	5.1 ± 0.6	8.1 ± 0.5
Gundruk	7.2 ± 0.2	<DL	7.4 ± 0.1
Inziangsang	7.2 ± 0.1	<DL	7.2 ± 0.2
Khalpi	7.4 ± 0.3	4.8 ± 0.5	7.6 ± 0.3
Sinki	7.2 ± 0.2	3.8 ± 0.2	7.6 ± 0.3
Ekung	7.8 ± 0.1	1.6 ± 0.2	7.9 ± 0.1
Eup	7.8 ± 0.1	<DL	7.9 ± 0.1
Hirring	7.9 ± 0.1	<DL	8.0 ± 0.1
Mesu	7.4 ± 0.2	3.7 ± 0.2	7.6 ± 0.3
Soibum	7.1 ± 0.3	<DL	7.2 ± 0.3
Soidon	6.0 ± 0.2	<DL	6.1 ± 0.2
Soijim	7.2 ± 0.2	<DL	7.2 ± 0.2

Note: Mold was not detected.

[a] LAB, lactic acid bacteria.
[b] AMC, aerobic mesophilic count.
[c] DL, less than detection limit (10 cfu/g).

and the identifications are interpreted using APILAB PLUS software (bio-Mérieux, France).

On the basis of a combination of phenotypic and genotypic identification methods, strains of LAB isolated from *gundruk, inziangsang, khalpi,* and *sinki* are identified as *Lactobacillus brevis, Lb. plantarum, Pediococcus pentosaceus,* and *P. acidilactici* (Tamang and Sarkar 1993; Tamang et al. 2005). The phenotypic characterization of LAB strains is followed by the randomly amplified polymorphic DNA (RAPD)-PCR technique (Tamang and Holzapfel 2004). The fingerprints obtained from this technique are used to identify clusters of closely related or even identical strains, allowing the selection of representative strains of these groups for further identification (Schillinger et al. 2003). A rep-PCR using (GTG)$_5$ primer is an appropriate tool to separate *Lb. plantarum* from the phenotypically almost indistinguishable *Lb. pentosus* and *Lb. paraplantarum,* and a species-specific PCR for *Lb. brevis* allowed the identification of the heterofermentative lactobacilli (Kostinek et al. 2005).

Genomic DNA is prepared from the selected strains, and fingerprinting analysis by the RAPD-PCR is conducted to reveal similarities among the strains of each group and the possible assignment of the strains to a certain species by including the respective reference strains in the assay. The strains of pediococci isolated from *gundruk* are grouped into one cluster at a similarity level above 75%, confirming them as *P. pentosaceus,* whereas the *Pediococcus* IB5 (*inziangsang*) showed a different profile, confirming it as *P. acidilactici* (Tamang et al. 2005). Based on the rep-PCR using the (GTG)$_5$ primer, all DAP-positive homofermentative strains (*gundruk, khalpi, inziangsang, mesu, soibum*) are confirmed as *Lb. plantarum* (Tamang and Holzapfel 2004). DAP-negative homofermentative strains isolated from *mesu* were genotypically identified as *Lb. curvatus* (Tamang et al. 2008). The strains of heterofermentative rods (from *khalpi* and *sinki*), phenotypically identified as the *Lb. brevis* group on the basis of sugar fermentation, were genotypically confirmed to be *Lactobacillus brevis* using the results of RAPD-PCR (Tamang et al. 2005).

On the basis of the RAPD profiles, the pediococci strains from *gundruk* and *inziangsang* are separated into *P. pentosaceus* and *P. acidilactici* (Tamang and Holzapfel 2004). The RAPD technique has been used to separate *P. pentosaceus* from *P. acidilactici*, isolated from Ethiopian fermented foods (Nigatu et al. 1998). *Lactobacillus plantarum, Lb. brevis, Lactococcus lactis, P. pentosaceus,* and *Enterococcus faecium,* isolated from *goyang,* and *Leuc. fallax* (*khalpi* and *sinki*) are identified on the basis of detailed phenotypic characteristics, including the API sugar profile data (Tamang 2006; Tamang and Tamang 2007). *Lactobacillus brevis, Lb. plantarum, P. pentosaceus,* and *P. acidilactici* have been reported to be involved in many lactic acid–fermented vegetable foods, including sauerkraut and cucumbers (Fleming et al. 1985; Font de Valdez et al. 1990; Tanasupawat et al. 1993; Randazzo et al. 2004). Enteroccoci play a beneficial role in the production of many fermented foods (Bouton et al. 1998; Cintas et al. 2000). *Enterococcus faecium* appears to pose a low risk for use in foods, because these strains generally harbor fewer recognized virulence determinants than *E. faecalis* (Franz et al. 2003). The presence of *Lactococcus lactis* in *goyang* supports an earlier report of its isolation from plant materials (Sandine et al. 1972), although the most recognized habitat for the lactococci is milk products (Teuber et al. 1991). There are not many reports on the occurrence of *Leuc. fallax* in foods. Strains of these species have been isolated from sauerkraut (Schillinger et al. 1989) and from *puto* (fermented rice cake of the Philippines) (Kelly et al. 1995). *Leuconostoc fallax* has been recognized as one of predominating organisms in the early heterofermentative stage of sauerkraut fermentation (Barrangou et al. 2002).

Lactobacillus plantarum, Lb. casei spp. *casei, Lb. casei* spp. *pseudoplantarum, Lb. fermentum,* and *P. pentosaceus* have been reported for *gundruk* of Nepal (Karki et al. 1983d). During fermentation of *gundruk, Lb. fermentum,* instead of *Leuconostoc mesenteroides* as in other fermented vegetable products, initiates the fermentation and is followed by homofermentative *P. pentosaceus* and finally *Lb. plantarum* (Dahal et al. 2005).

Strains of LAB isolated from *mesu, soidon, soibum,* and *soijim* are identified using phenotypic tests and genotypic methods, such as rep-PCR and RAPD-PCR, as *Lb. brevis, Lb. plantarum, Lb. curvatus,* and *P. pentosaceus*

(Tamang and Sarkar 1996; Tamang et al. 2008). For the other LAB strains—
Leuc. mesenteroides, *Leuc. fallax*, *Leuc. lactis*, *Leuc. citreum*, and *Enterococcus durans*—and also strains from *ekung, eup,* and *hirring* of Arunachal Pradesh, phenotypic characterization including DAP, lactate configuration, and API system is performed for the identification to species level (Tamang and Tamang 2009). The typical enterococci (*Enterococcus durans*) can be easily distinguished from other Gram-positive, catalase-negative, homo-fermentative cocci such as streptococci and lactococci by their ability to grow both at 10°C and 45°C in 6.5% NaCl and at pH 9.6 (Franz et al. 2003). *Lactobacillus casei* has been found as the dominant bacterium in naturally fermented Sicilian green olives (Randazzo et al. 2004). *Leuconostoc citreum* is reported in traditional French cheese (Cibik et al. 2000). Recovery of *Lb. plantarum, Lb. brevis, Leuc. fallax, Leuc. mesenteroides,* and *E. durans* in *soibum* of Manipur has updated the earlier findings of Giri and Janmejay (1994), the identification of which was based on limited phenotypic characteristics (Giri and Janmejay 1987). Similar fermented bamboo-shoot products called *naw-mai-dong* or *nor-mai-dorng* of Thailand also contain lactobacilli, leuconostocs, and pediococci (Dhavises 1972; Phithakpol et al. 1995). Although, *E. durans* has been isolated from Feta cheese (Igoumenidou et al. 2005), there is no published report of its presence in any fermented vegetable product. Based on the characterizations and identification profiles of LAB strains isolated from ethnic fermented bamboo products of Arunachal Pradesh, species of LAB are identified from *ekung* samples (*Lb. plantarum, Lb. brevis, Lb. casei,* and *Tetragenococcu halophilus*), from samples of *eup* (*Lb. plantarum, Lb. fermentum*), and from *hiring* (*Lb. plantarum, Lc. lactis*) (Tamang and Tamang 2009).

Though the dominant microorganisms in the Himalayan fermented vegetable and bamboo products are lactic acid bacteria, a sizable number of yeasts, mostly *Pichia, Candida, Saccharomyces,* and *Rhodotorula* species, are also reported in some finished products of *goyang, khalpi, sinki, ekung, hirring,* and *mesu* in numbers ranging between 10^4 and 10^6 cfu/g (Tamang 2006). These yeasts might be spoilage in the products, or they might have appeared during storage. Many genera and species of yeasts occur in *kimchi* (Song and Park 1992). *Hansenula, Saccharomyces,* and *Torulopsis* species are reported from cucumber pickles (Etchells et al. 1961). Some oxidative yeasts belonging to *Candida, Pichia, Debaryomyces, Saccharomyces,* and *Hansenula* species are associated with gassy fermentation and softening of olives (Vaughn et al. 1972).

About 65% of LAB strains isolated from the Himalayan fermented vegetable and bamboo shoots belong to *Lactobacillus*. Homofermenter lactobacilli are represented by 45.4%, followed by heterofermenter lactobacilli representing 20%; and leuconostoc 16.3%, pediococci 12.7%, lactococci 1.7%, and *Tetragenococcus* 0.7% of total microflora of the Himalayan fermented vegetable products (Tamang 2006; Tamang and Tamang 2008).

2.2.1   Food safety

The occurrence of bacterial contaminants—mostly *Staphyloccus aureus* in a few finished samples of *goyang, gundruk,* and *sinki; Bacillus cereus* in a few dry samples of *sinki;* and Enterobacteriaceae in *goyang* and *gundruk* samples—has been observed at levels of around 10^2 cfu/g (Tamang 2006). These pathogenic bacteria might have been introduced during handling of raw materials for preparation, when pH is not low enough to inhibit their growth. Otherwise, no pathogenic bacteria such as *Listeria, Salmonella,* and *Shigella* species have been detected in the Himalayan fermented vegetables and bamboo products due to the acidic nature of the products. A small number of *Bacillus cereus* in foods is not considered significant (Roberts et al. 1996). Rapid growth of LAB could restrict the growth of other organisms simply by their physical occupation of available space and uptake of the most readily assimilative nutrients (Adams and Nicolaides 1997). Moreover, LAB may reduce pH to a level where pathogenic bacteria may be inhibited or destroyed (Holzapfel et al. 1995; Tsai and Ingham 1997).

2.2.2   Functional or technological properties

The technological properties of LAB are necessary criteria for selection of potent starter cultures among the LAB for upgrading the traditional process and development of functional foods (Durlu-Ozkaya et al. 2001; Badis et al. 2004). Acidification is an important technological property with relevance in the selection of a starter culture among the LAB (de Vuyst 2000). LAB strains isolated from the Himalayan fermented vegetable and bamboo-shoot products were screened for their acidifying and coagulating capacities, and it was found that most of the LAB strains acidified the products by lowering the pH to 4.0 in addition to showing strong coagulating activities (Tamang 2006). The ability of some species of LAB, particularly *Lb. plantarum,* in acidification of the substrates is significant in food preservation (Ammor and Mayo 2007). These strains (isolated from the Himalayan fermented vegetable foods), although originating from plant sources, appeared to be adapted to the milk ecology, since they coagulated and acidified the skim milk used in the applied method (Tamang 2006). Justifying milk coagulation by the LAB strains isolated from fermented vegetables, the enzymatic activity of LAB strains show high activity of peptidases, which may be responsible for coagulation of skim milk in the applied method (Tamang 2006). The majority of strains of LAB grow well in bile salt, showing their ability to tolerate bile (Tamang and Tamang 2009). Bile-salt tolerance is considered an important colonization factor for probiotic bacteria (du Toit et al. 1998).

Strains of different species of *Lactobacillus, Pediococcus,* and *Leuconostoc* show antimicrobial activities against a number of potentially pathogenic

Gram-negative and Gram-positive bacteria and simultaneously perform an essential role in the preservation of perishable vegetables (Schillinger et al. 2004). LAB compete with other microbes by screening antagonistic compounds and modifying the microenvironment by their metabolism (Lindgren and Dobrogosz 1990). It is interesting to note that only one strain of *Lb. plantarum*, IB2, isolated from *inziangsang*, fermented vegetable of Nagaland and Manipur, was found to produce a bacteriocin against *Staphylococcus aureus* S1 (Tamang and Holzapfel 2004). This may be explained by the fact that other factors are also involved in antagonism, and by the influence of growth rate and competitiveness of a culture for bacteriocin production (Tagg 1992; Yang and Ray 1994). This is determined by its adaptation to a substrate and by a number of intrinsic and extrinsic factors, including redox potential, water activity, pH, and temperature (Holzapfel et al. 1995; Ouwehand 1998). A number of Gram-positive pathogenic bacteria, including *Staphylococcus aureus*, have been found sensitive to bacteriocin of *Lactobacillus* spp. (Tichaezek et al. 1992; Niku-Paavola et al. 1999). Our finding on bacteriocin activity of *Lb. plantarum* IB2 (*inziangsang*) against *Staphylococcus aureus* is justified by the recent report on the inhibitory effect of bacteriocin produced by *Lb. plantarum* in foods against *Staphylococcus aureus* and other Gram-negative bacteria (Jamuna et al. 2005). Crude bacteriocin activity of *Lb. plantarum* IB2 is quantified and calculated as 32 AU/ml (Tamang 2006). Quantification is necessary to observe the quotient between the inhibition zone area and the sensitivity of the indicator bacterium used to compare different bacteriocins (Delgado et al. 2005). Species of LAB strains isolated from several fermented vegetable products have the antimicrobial activities, including bacteriocins and nisin production, such as fermented olives (Rubia-Soria et al. 2006), sauerkraut (Tolonen et al. 2004), fermented carrots (Uhlman et al. 1992), fermented cucumbers (Daeschel and Fleming 1987), and organic leafy vegetables (Ponce et al. 2008).

The absence of proteinases (trypsin and chymotrypsin) and the presence of strong peptidase (leucine arylamidase, valine arylamidase, and cystine arylamidase) and esterase-lipase (C4 and C8) activities produced by the predominant LAB strains isolated from the Himalayan fermented vegetable products are possible traits of desirable quality for their use in production of typical flavor (Tamang and Holzafpel 2004).

Antinutritive factors such as phytic acids and oligosaccharides are of particular significance in unbalanced cereal-based diets (Holzapfel 2002; Fredrikson et al. 2002). Oligosaccharides such as raffinose, stachyose, and verbascose cause flatulence, diarrhea, and indigestion (Abdel Gawad 1993; Holzapfel 1997). Due to these nutritional consequences, the degradation of antinutritive factors in food products by fermentation is desirable, as reported for a number of foods of plant origin (Chavan and Kadam 1989;

Mbugua et al. 1992). High activity of phosphatase by LAB strains shows their role in phytic acid degradation in the Himalayan fermented vegetable products (Tamang and Holzapfel 2004). About 77% of *Lb. plantarum* strains isolated from the Himalayan fermented vegetable and fermented bamboo-shoot products is found to degrade raffinose, while none of the *P. pentosaceus* strains and *P. acidilactici* are able to degrade raffinose (Tamang 2006). This is in agreement with the report of Holzapfel (2002), where *Lb. plantarum* strains isolated from fermented Ghanaian maize products are able to ferment raffinose, while strains of *P. pentosaceus* and *P. acidilactici* are unable to do so.

Biogenic amines are organic basic compounds that occur in different kinds of foods such as sauerkraut (Taylor et al. 1978), fishery products, cheese, wine, beer, dry sausages, and other fermented foods (ten Brink et al. 1990), fruits, and vegetables (Suzzi and Gardini 2003). Histaminine, a precursor of biogenic amine, has been recognized as the causative agent of scombroid poisoning (histamine intoxication), whereas tyramine has been related to food-induced migraines and hypertension (Taylor 1986; Bover-cid and Holzapfel 1999). Leafy vegetables usually contain low levels of biogenic amines, but these may increase during fermentation due to the decarboxylase activity of microorganisms (Silla-Santos 2001). It is found that LAB strains isolated from the Himalayan fermented vegetables do not produce biogenic amines, which is a good indication of their potential for possible development as starter culture (Tamang and Tamang 2009). The production of biogenic amines is not a desirable property for LAB selected as starter cultures (Buchenhüskes 1993; Holzapfel 1997).

Bacterial adherence to hydrocarbons, such as hexadecane, proved to be a simple and rapid method to determine cell surface hydrophobicity (Rosenberg et al. 1980; van Loosdrecht et al. 1987; Ding and Lämmler 1992; Vinderola et al. 2004). A few strains of LAB isolated from the Himalayan fermented vegetable products show more than 75% hydrophobicity, indicating their hydrophobic nature (Tamang 2006; Tamang and Tamang 2009). A hydrophobicity of greater than 70% is arbitrarily classified as hydrophobic (Nostro et al. 2004). The high degree of hydrophobicity of the LAB strains, isolated from the Himalayan fermented vegetable products, probably indicates the potential of adhesion to the gut epithelial cells of the human intestine, thereby advocating their probiotic character (Holzapfel et al. 1997, 1998), provided that these strains are consumed in a viable state. The ability of LAB to adhere to the intestinal mucosa is considered one of the main criteria in the selection of a potential probiotic culture (Shah 2001; Holzapfel and Schillinger 2002). The functional effects of probiotic bacteria include adherence to the intestinal cell wall for colonization in the gastrointestinal tract (GIT), with a capacity to prevent pathogenic adherence or pathogen activation (Spinler et al. 2008).

It has been observed that, within the same species, each strain of LAB that was isolated from the Himalayan fermented vegetable and bamboo-shoot products responds to different characteristics in regard to the technological or functional properties. This differentiation may lead to a diversity of different strains within the same species that would potentially have different technological or health-promoting benefits in the products.

2.2.3   In situ fermentation dynamics

In situ fermentation or succession change in microorganism(s) during natural fermentation is one of the important criteria in the study of fermented foods to understand the composition of the functional microorganisms and their effects on the physicochemical changes of the fermenting substrates. Important physicochemical aspects during fermentation from 0 h to the end of fermentation include pH, incubating temperature of the fermenting substrates, acidity, moisture, ash, reducing sugars, total nitrogen, etc. Microbial studies include the decrease/increase in population size of total viable counts, LAB, yeasts, molds, pathogenic bacteria, etc.

During natural fermentation of *gundruk*, indigenous lactic acid bacteria change spontaneously, and at the end of the fermentation, *Lactobacillus* species, mainly *Lb. plantarum*, is involved (Tamang and Tamang 2008). Spontaneous changes in LAB populations during fermentation of several vegetables involving lactobacilli have been reported (Fernandez Gonzalez et al. 1993; Harris 1998). As expected in a typical lactic acid fermentation (Vaughn 1985; Lee 1997; Lu et al. 2001), the pH of the fermenting substrates decreases, and the titratable acidity increases as the *gundruk* fermentation progresses due to growth of LAB, which converts fermentable sugars into lactic acid. *Saccharomyces* sp., *Pichia* sp., and *Zygosaccharomyces* sp. are found during the initial stage of *gundruk* fermentation and then disappear as the fermentation progresses (Tamang and Tamang 2008). Similarly, the population of pathogenic contaminants disappears during fermentation because of the dominance of LAB. By averting the invasion of these contaminants, lactic acid fermentation imparts safety to a product like *gundruk*. There has been no report of any food poisoning or infectious disease infestation from the consumption of fermented vegetables or fermented bamboo shoots in the Himalayas. The LAB produce sufficient acid for inhibition of pathogenic microorganisms in foods (Nout 1994; Adam and Nicolaides 1997). *Gundruk* fermentation is initiated by *Lb. fermentum* and is followed by *Pediococcus pentosaceus* and finally by *Lb. plantarum, Lb. casei*, and *Lb. casei* subsp. *pseudoplantarum* (Karki et al. 1983d; Tamang and Holzapfel 2004). These bacteria produce lactic acid and acetic acid, which lower the pH of the substrates, making the products more acidic in nature (Dietz 1984; Karki et al. 1983d). Due to low pH (3.3–3.8) and high acid content (1.0%–1.3%), *gundruk* and *sinki* can be preserved for longer periods without refrigeration. This is cited as a practical example of biopreservation of perishable vegetables in the Himalayan regions.

The difference in incubation temperature of the fermenting cucumbers during *khalpi* fermentation is not significant ($p < .05$) (Tamang and Tamang 2008). The pH decreases significantly ($p < .05$) from 5.6 to 3.2, and the percent acidity increases significantly ($p < .05$) from 0.28% to 1.24% at the end of fermentation. The population of LAB in raw cucumber is very small (10^3 cfu/g), which increases significantly ($p < .05$) to 10^8 cfu/g within 36 h, and then remains at the level of 10^7 cfu/g in the final product. The predominant LAB in *khalpi* fermentations are *Leuc. fallax*, *P. pentosaceus*, *Lb. brevis*, and *Lb. plantarum* (Tamang et al. 2005). *Leuconostoc* is the major bacterial genus in the initial phase of lactate fermentation of vegetables (Eom et al. 2007). Heterofermentative LAB such as *Leuc. fallax*, *Lb. brevis*, and *P. pentosaceus* are isolated from the initial fermentation stage of *khalpi*. As the fermentation progresses, more acid-producing homofermentative lactobacilli, mainly *Lb. plantarum*, remain dominant. The load of *Staphylococcus aureus* and Enterobacteriaceae is reduced significantly ($p < .05$) and eventually disappears during fermentation. Acidity, pH, and buffer capacity greatly influence the establishment and extent of growth of LAB during cucumber fermentation (McDonald et al. 1991). The microbial load of yeast in raw cucumber is 10^4 cfu/g, which disappears after 48 h. Yeasts detected during the initial stage of *khalpi* fermentation are *Pichia* sp., *Candida* sp., and *Saccharomyces* sp. (Tamang and Tamang 2008). Fresh vegetables, like cucumbers, contain a numerous and varied epiphytic microflora, including many spoilage microorganisms, and an extremely small population of LAB (Mundt and Hammer 1968). Numerous chemical and physical factors influence the growth of various microorganisms, as well as their sequence of appearance during the fermentation (Fleming and McFeeters 1981).

2.2.4   *Upgrading of traditional process*

Vegetable fermentation processes in the Himalayas vary from place to place, causing inconsistency in the final product. Moreover, the natural fermentation sometimes results in inferior quality and unacceptability of the product due to fluctuating incubation temperatures, the mix of microorganisms, environmental conditions, etc. An attempt was made to upgrade the traditional processing of perishable vegetables using pure strains of lactic acid bacteria isolated from the traditionally fermented vegetable products of the Himalayas (Tamang and Tamang 2009). *Gundruk* was selected because it is widely consumed in the various regions of the Himalayas, and *khalpi* was selected because cucumbers are in plentiful supply.

On the basis of superior technological properties of LAB strains (Tamang 2006), such as acidification ability, antimicrobial activities, nonproduction of biogenic amines, ability to degrade antinutritive factors, and even a high degree of hydrophobicity, *Lb. plantarum* GLn:R1 (MTCC 9483) and *P. pentosaceus* GLn:R2 (MTCC 9484) were selected as a starter culture for production of *gundruk*. (Cultures were obtained from Microbial Type Culture Collection [MTCC] of the Institute of Microbial Technology, Chandigarh, India.)

Preparation of *gundruk* using *Lb. plantarum* GLn:R1and *P. pentosaceus* GLn:R2 proceeded as follows: Leaves of local vegetable *rayo-sag* (*Brassica rapa* L. subsp. *campestris* [L.] Clapham var. *cumifolia* Roxb.; family Brassicacea) were washed thoroughly in sterile distilled water and wilted in oven (≈30°C) for 6 h. Leaves were crushed, put into sterile warm water (about 90°C) for 5 min, and transferred into another sterile glass container. Excess water in the leaves was removed by squeezing, and about 400 g of crushed leaves were distributed aseptically into each sterile 500-ml capped bottle. Each bottle was inoculated with a mixture of actively grown culture strains of *Lactobacillus plantarum* GLn:R1 (MTCC 9483) and *Pediococcus pentosaceus* GLn:R2 (MTCC 9484), previously isolated from market samples of *gundruk* (Tamang et al. 2005), at the ratio of 10^7 cfu/g. Bottles were tightly capped and incubated at 20°C, 25°C, or 30°C for 6 days. *Gundruk* prepared by the starter culture was evaluated organoleptically. With respect to general acceptability, the 6-day-old *gundruk* fermented at 20°C had the highest score, with better aroma and an acidic taste typical of *gundruk* that is acceptable to consumers (Tamang and Tamang 2009).

Preparation of *khalpi* using *Lb. plantarum* KG:B1, *Lb. brevis* KG:B2, and *Leuc. fallax* KB:C1 was carried out as follows: Ripened cucumber (*Cucumis sativus* L.) was cleaned, washed, and cut into pieces and oven dried at ≈30°C for 8 h. About 400 g of oven-dried pieces of cucumber were transferred into each sterile 250-ml bottle. *Lactobacillus plantarum* KG:B1 (MTCC 9485), *Lb. brevis* KG:B2 (MTCC 9486), and *Leuc. fallax* KB:C1 (MTCC 9487), previously isolated from *khalpi* (Tamang et al. 2005), were selected on the basis of superior technological properties such as acidifying capacity, antimicrobial activities, nonproduction of biogenic amines, and ability to degrade antinutritive factors of the raw materials (Tamang 2006). Each bottle was inoculated with a mixture of pure culture strains of actively grown *Lb. plantarum* KG:B1 (MTCC 9485), *Lb. brevis* KG:B2 (MTCC 9486), and *Leuc. fallax* KB:C1 (MTCC 9487) at the ratio of 10^7 cfu/g. Bottles were tightly capped and incubated at 20°C, 25°C, and 30°C for 72 h, and samples were tested for sensory evaluation (Tamang and Tamang 2009). *Khalpi* produced with a mixed pure culture at 20°C for 72 h had the highest score in aroma (Tamang and Tamang 2009). There was a significant ($p < .05$) increase in taste score with time at 20°C, with a high score of 4.5 at 72 h. The highest score ($p < .05$) in taste (strongly acidic) was observed at 72 h for product fermented at 30°C. The general acceptability score was highest in *khalpi* fermented at 20°C for 72 h. *Khalpi* produced at 25°C was also more or less similar within 48 h, while at 30°C a strongly acidic taste developed quickly and was not preferred by the consumers (Tamang and Tamang 2009).

Gundruk prepared by using a starter culture has advantages over the conventional method because of the shorter fermentation time and a greater consistency in quality and flavor. Similarly, the use of pure culture strains in *khalpi* preparation took less time; under natural conditions, it takes more than 3 days for fermentation to complete. The use of mixed starter culture complements different technological properties to attain better products (Oguntoyinbo et al. 2007). The use of starter culture in the production of fermented foods increases the safety of processes and reduces losses caused by false fermentation (Geisen and Holzapfel 1996). Authenticating the identity of functional microbes in fermented foods is necessary to develop the starter cultures isolated from conventionally pre-pared foods (Tamang and Holzapfel 1999). An optimized process condition is always superior to the conventional method; however, the introduction and replacement of natural and easily operated traditional technologies may be difficult to change for the rural producers of the Himalayas.

2.3 Nutritive value

The moisture content in *gundruk, sinki, inziangsang,* and *eup* is low due to drying after fermentation; other fermented vegetable and bamboo prod-ucts of the Himalayas are freshly prepared with high moisture content (Tamang 2006). The Himalayan fermented vegetable and bamboo-shoot foods have comparatively higher mineral content (Tables 2.2 and 2.3). Higher mineral contents are observed in Indian vegetables (Gopalan et al. 1995).

The food values of fermented vegetable and bamboo-shoot products of the Himalayas are almost the same as reported in other fermented vegetable products, such as *kimchi* (Cheigh and Park 1994), sauerkraut (Campbell-Platt 1987), and fermented cucumber (Fleming 1984). The pH and acidity (as lactic acid) in *gundruk* are 4.0–4.3 and 0.8%–1.0%, respec-tively (Karki et al. 1983d). In *gundruk,* 90% of the organic acids consist of lactic and acetic acids, and other organic acids are citric, malic, and acetic acids (Karki 1986). The level of palmitic, oleic, linoleic, and linolenic acids is much higher in mustard leaf *gundruk* compared to those in the unfer-mented vegetables (Karki et al. 1983c). In mustard *gundruk,* free amino acids—mostly glutamic acid, alanine, leucine, lysine, and threonine—remarkably increase with a decrease in asparagine, glutamine, histidine, and arginine during fermentation (Karki et al. 1983b). *Gundruk* prepared from leaves of mustard comprises cyanides (15.7%) and isothiocyanates (8.5%) as the main flavor components, followed by alcohols (12.3%), esters

Table 2.2 Nutritional Composition of Himalayan Fermented Vegetable Products

Parameter[a]	Product				
	Goyang	Gundruk	Inziangsang	Khalpi	Sinki
Moisture %	92.5 ± 3.0	15.0 ± 1.8	17.6 ± 0.8	91.4 ± 1.0	22.8 ± 1.2
pH	6.5 ± 0.1	5.0 ± 0.1	4.8 ± 0.2	3.9 ± 0.2	4.1 ± 0.1
Titratable acidity (% as lactic acid)	0.13 ± 0.04	0.49 ± 0.12	0.50 ± 0.07	0.95 ± 0.09	0.65 ± 0.1
Ash (% DM)	12.9 ± 1.7	22.2 ± 1.4	16.9 ± 0.3	14.2 ± 2.2	15.6 ± 0.9
Fat (% DM)	2.1 ± 0.1	2.1 ± 0.2	3.2 ± 0.6	2.6 ± 0.5	1.4 ± 0.2
Protein (% DM)	35.9 ± 1.5	37.4 ± 0.9	38.7 ± 0.8	12.3 ± 1.6	14.9 ± 0.7
Carbohydrate (% DM)	48.9 ± 2.8	38.3 ± 1.9	41.2 ± 2.2	70.9 ± 3.6	68.0 ± 2.1
Food value (kcal/100 g DM)	357.2 ± 9.9	321.9 ± 3.5	348.4 ± 2	356.2 ± 1.5	344.2 ± 1.6
Minerals (mg/100 g)					
Ca	92.2 ± 2.0	234.6 ± 5.2	240.4 ± 3.6	6.4 ± 0.3	223.9 ± 0.7
Na	6.7 ± 1.1	142.2 ± 2.0	133.7 ± 4.2	2.2 ± 0.2	737.3 ± 2.6
K	268.4 ± 6.6	677.6 ± 6.8	658.4 ± 6.7	125.1 ± 2.3	2320.4 ± 4.8

Note: Data represent the means (± SD) of triplicate of each sample.

[a] DM, dry matter.

Table 2.3 Nutritional Composition of the Himalayan Fermented Bamboo-Shoot Products

Parameter	Product					
	Ekung	Eup	Hirring	Mesu	Soibum	Soidon
Moisture %	94.7 ± 0.7	36.8 ± 0.2	88.8 ± 0.5	89.9 ± 0.2	92.0 ± 0.4	92.2 ± 0.5
pH	3.9 ± 0.4	4.1 ± 0.2	4.0 ± 0.2	3.9 ± 0.2	3.9 ± 0.1	4.2 ± 0.1
Titratable acidity (% as lactic acid)	0.94 ± 0.1	0.80 ± 0.2	0.81 ± 0.2	0.88 ± 0.1	0.98 ± 0.3	0.96 ± 0.1
Ash (% DM)	14.0 ± 1.3	18.2 ± 1.8	15.0 ± 1.2	15.0 ± 1.3	13.3 ± 0.2	13.1 ± 4.0
Fat (% DM)	3.8 ± 1.7	3.1 ± 0.1	2.7 ± 1.1	2.6 ± 0.8	3.2 ± 0.2	3.1 ± 0.2
Protein (% DM)	30.1 ± 1.7	33.6 ± 4.0	33.0 ± 4.4	27.0 ± 2.7	36.3 ± 0.3	37.2 ± 0.3
Carbohydrate (% DM)	52.1 ± 6.2	45.1 ± 4.2	49.3 ± 8.8	55.6 ± 4.8	47.2 ± 0.4	46.6 ± 0.4
Food value (kcal /100g DM)	363.0 ± 5.8	342.7 ± 6.3	353.5 ± 6.5	352.4 ± 6.3	362.8 ± 4.5	363.1 ± 4.6
Minerals (mg/100 g)						
Ca	35.4 ± 6.5	76.9 ± 9.3	19.3 ± 6.1	7.9 ± 0.4	16.0 ± 0.7	18.5 ± 0.9
Na	10.9 ± 3.2	3.4 ± 0.7	3.4 ± 0.8	2.8 ± 1.1	2.9 ± 0.9	3.7 ± 0.7
K	168.6 ± 7.2	1815.6 ± 7.5	272.4 ± 6.8	282.6 ± 3.2	212.1 ± 0.9	245.5 ± 0.8

Note: Data represent the means (± SD) of triplicate of each sample.

[a] DM, dry matter.

(4.1%) and phenyl acetaldehyde (6.4%) (Karki et al. 1983a). *Gundruk* made from cauliflower leaves contains alcohols (50%) as the major component followed by cyanides (6.5%), isothiocyanates (6.1%), and esters (3.2%) (Karki et al. 1983a). Proline content in mustard vegetable is greater than that in cauliflower *gundruk* or mustard *gundruk*. This may be due to the wilting of vegetables prior to fermentation (Karki et al. 1983b). The levels of iron and calcium are high, while carotenoids are reduced by more than 90%, probably during sun drying (Diez 1984).

Due to the high content of organic acids, *gundruk* and *sinki* are considered to be good appetizers. During fermentation of *soibum*, a degradation of protein and an increase in free amino acids have been observed (Giri and Janmejay 1994, 2000). Enhancement of nutritional value in *soibum* has been reported by Pravabati and Singh (1986). *Bacillus subtilis, B. licheniformis, B. coagulans,* and *Micrococcus luteus* isolated from *soibum* were found to be involved in microbial bioconversion of phytosterols (Sarangthem and Singh 2003). *Lung-siej* contains higher levels of iron and magnesium than raw bamboo (Agrahar-Murugkar and Subbulaksmi 2006).

2.4   Conclusion

Scientific knowledge on the Himalayan ethnic fermented vegetables is sparse outside the region. The predominant LAB species associated with biopreservation of perishable vegetables and bamboo shoots are summarized as follows: *Lb. plantarum, P. pentosaceus* (*gundruk*); *Lb. plantarum, Lb. brevis, Lb. casei, Leuc. fallax* (*sinki*); *Lb. plantarum, Lb. brevis, Lc. lactis, E. faecium, P. pentosaceus* (*goyang*); *Lb. plantarum, Lb. brevis, P. acidilactici* (*inziangsang*); *Lb. plantarum, Lb. brevis, Leuc. fallax* (*khalpi*); *Lb. plantarum, Lb. brevis, Lb. curvatus, Leuc. citreum, P. pentosaceus* (*mesu*); *Lb. plantarum, Lb. brevis, Leuc. fallax, Leuc. lactis, Leuc. mesenteroides, E. durans* (*soibum*); *Lb. brevis, Leuc. fallax, Leuc. lactis* (*soidon*); *Lb. brevis, Leuc. fallax, Leuc. lactis* (*soijim*); *Lb. plantarum, Lb. brevis, Lb. casei, Tetragenococcus halophilus* (*ekung*); *Lb. plantarum, Lb. fermentum* (*eup*); and *Lb. plantarum, Lc. lactis* (*hirring*). Strains of LAB play a complex role in the traditional fermentation by their functional properties related to a specific enzyme spectrum, acidifying capacity, degradation of antinutritive factors, antimicrobial activities, probiotic properties, and even as nonproducers of biogenic amines. Some strains of LAB may be used as starter culture for controlled and optimized production of the Himalayan fermented vegetable and bamboo shoots.

chapter three

Fermented legumes

Several indigenous varieties of legumes are grown in the Himalayas, including soybeans, black grams, beans, etc. Soybean (*Glycine max* [L.] Merrill, family Leguminosae, subfamily Papilionaceae) is a summer leguminous crop, grown under rain-fed conditions in upland terraces as a sole crop as well as a mixed crop with rice and maize up to an elevation of 1500 m. Soybean, locally known as *bhatmas* in the Nepali language, is traditionally used to prepare various fermented and nonfermented recipes in the eastern Himalayan regions of Nepal, India, and Bhutan (Tamang 1996). Two indigenous varieties of soybeans—yellow cultivar and dark brown cultivar—are grown in between May and June and harvested in November. Locally grown soybeans are harvested, and dry seeds of soybeans are naturally fermented into a flavorsome and sticky product in the eastern parts of Nepal, Darjeeling hills, Sikkim, northeastern regions of India, and southern parts of Bhutan by the Mongolian races. This fermentation process is primarily for improvement of sensory quality and nutritional value, rather than preservation.

3.1 Important fermented soybean foods

Some of the common ethnic nonsalted sticky fermented soybean foods of the eastern Himalayas are *kinema* (Nepal, Darjeeling hills, Sikkim, and South Bhutan), *hawaijar* (Manipur), *tungrymbai* (Meghalaya), *bekang* (Mizoram), *aakhone* (Nagaland), *peruyyan* (Arunachal Pradesh), and a fermented black gram condiment known as *maseura* (Nepal, Darjeeling hills, and Sikkim). It is interesting to note that no soy sauce–like product, or miso-like paste, or mold-fermented tempeh, or *tufu*-like soybean curd and its fermented product *sufu* (Han et al. 2001) is traditionally prepared in the eastern Himalayas. Fermentation of soybean is limited only to sticky, alkaline, and flavorsome foods in the eastern Himalayas, including North East India. With rapid globalization in the cities and towns, the use of imported or commercial soy sauce is becoming popular among the urban populace in the Himalayas.

3.1.1 Kinema

Kinema (Figure 3.1) is an indigenous fermented soybean food that is a sticky, slightly alkaline product with a slight ammoniacal flavor produced by natural fermentation. *Kinema* is an inexpensive and high-protein

Figure 3.1 Fresh *kinema* of Darjeeling hills.

plant food in the local diet. It is a fermented and flavorsome whole-soybean food with a sticky texture and a gray–tan color (Tamang 2001b).

3.1.1.1   Indigenous knowledge of preparation

During traditional production of *kinema* in Darjeeling hills, Sikkim, and South Bhutan, the small (≈6 mm) yellow-cultivar soybean dry seeds are selected, washed, and soaked overnight (8–10 h) in water. Soaked soybean seeds are taken out and put into a container with fresh water and boiled for 2–3 h until they are soft. Excess water is drained off, and the cooked soybean seeds are filled into a wooden mortar, locally called *okhli*, and are cracked lightly by a wooden pestle, locally called *muslo*, to split the cotyledons. This practice of cracking cooked seeds of soybeans is observed only during *kinema* production, unlike other similar fermented soybean foods of Asia and North East India, probably to increase the surface areas and speed the fermentation by aerobic spore-forming *Bacillus* spp. About 1% of firewood ash is added directly to the cooked soybeans and mixed thoroughly to maintain the alkaline condition of the product. Soybean grits are placed in a bamboo basket lined with locally grown fresh fern (*Glaphylopteriolopsis erubescens*). The basket is covered with a jute bag and left to ferment naturally at ambient temperatures (25°C–40°C) for 1–3 days above an earthen oven in the kitchen (Figure 3.2). During summer, the fermentation time may require 1–2 days, while in winter it may require 2–3 days (Tamang 1994).

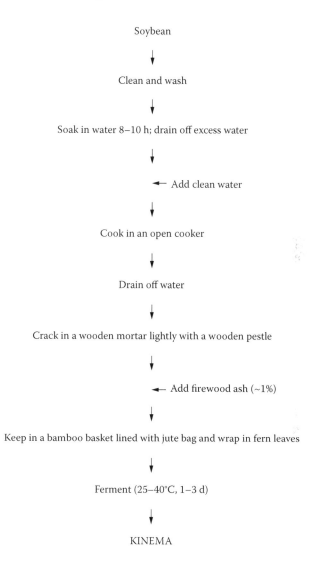

Soybean

↓

Clean and wash

↓

Soak in water 8–10 h; drain off excess water

↓

◄— Add clean water

↓

Cook in an open cooker

↓

Drain off water

↓

Crack in a wooden mortar lightly with a wooden pestle

↓

◄— Add firewood ash (~1%)

↓

Keep in a bamboo basket lined with jute bag and wrap in fern leaves

↓

Ferment (25–40°C, 1–3 d)

↓

KINEMA

Figure 3.2 Indigenous method of *kinema* production in Nepal and Sikkim.

In eastern Nepal, local consumers prepare dark brown local varieties of soybean seeds rather than yellow-colored seeds for making *kinema*
(Nikkuni 2007). Similarly, they commonly use *Ficus* (fig) and banana leaves
as wrapping materials instead of fern fronds. Other methods remain the
same. Completion of fermentation is indicated by the appearance of a white
viscous mass on the soybean seeds and the typical *kinema* flavor, with a
slight odor of ammonia.

The shelf life of freshly prepared *kinema* remains 2–3 days in summer and a maximum of a week in winter without refrigeration. It can be prolonged by drying in the sun for 2–3 days. Dried *kinema* is stored for several months at room temperature. Preparation of *kinema* varies from place to place and is still restricted to the household level. It is interesting to note that the mountain women use their indigenous knowledge of food production prepare *kinema*. This unique knowledge of *kinema* making has been protected as a hereditary right and is passed from mother to daughter, mostly among the Limboo.

Kinema is similar to other Asian *Bacillus*-fermented sticky soybean foods, such as *natto* of Japan (Hosoi and Kiuchi 2003), *chungkukjang* of Korea (Shon et al. 2007), *thua nao* of northern Thailand (Inatsu et al. 2006), *pepok* of northern Myanmar (Tanaka 2008a), and *sieng* of Cambodia (Tanaka 2008b).

The preparation of *kinema* is very similar to that of *natto*. In *itohiki-natto*, whole soybeans are used for fermentation, and in *hikiwari-natto*, dehulled soybeans cracked into two to four pieces are used (Ohta 1986). Some of the steps in *kinema* preparation do not resemble those in *natto*, and thus make *kinema* a unique nonsalted soybean fermented product. The cooked beans are lightly crushed to dehull most of the seeds, but fermentation is carried out with the kernels as well as the seed coats. While *natto* is consumed as it is with *shoyu* (Kiuchi et al. 1976), *kinema* is always fried in oil and made into a curry. The practice of frying *kinema* may have developed to drive out the unpleasant ammonia smell, which masks the pleasant and persistent nutty flavor.

3.1.1.2   Culinary practices

Kinema is eaten as a curry with steamed rice (Figure 3.3). The delicacy of *kinema* can be perceived by its appealing flavor and sticky texture. Fresh *kinema* is fried in vegetable oil with chopped onions, tomatoes, and turmeric powder. Salt and sliced green chillies are added and fried for 3–5 min. A little water is added to make a thick gravy and cooked for 5–7 min, and then the *kinema* curry is ready for serving with steamed rice (Tamang and Tamang 1998). Dried *kinema* is sometimes mixed with leafy vegetables to make a mixed curry.

Consumers like *kinema* mostly due to its typical flavor and sticky texture, although some people dislike the product due to its strong umami-type flavor and mucilaginous texture.

3.1.1.3   Economy

Kinema production is a source of marginal income generation for many families in the eastern Himalayas. *Kinema* is sold by rural women in all local periodical markets, called *haats* in eastern Nepal, Darjeeling hills, Sikkim, and southern parts of Bhutan (Figure 3.4). Usually, it is sold by volume, measured in a small silver mug containing 150–200 g of *kinema* and

Figure 3.3 *Kinema* curry.

Figure 3.4 The Limboo woman selling *kinema* at Gangtok market using a silver mug for weighing.

packed in the leaves of a fig plant (*Ficus hookeriana*), locally called *nevara*, and then tied loosely by straw. Poly-bags are not used to pack *kinema*. One kg of *kinema* costs about Rs. 50. An average of 5 kg of *kinema* is sold by each seller in a local market, and about 60% of the expenses are incurred for the purchase of raw soybeans, fuel for cooking, transportation from village to market, etc., for a net profit of 40% (Tamang 1998b). This profit is spent on children's education, procuring essential commodities not locally available, and other domestic expenses.

Although there is a good demand for *kinema* in the local markets, production of *kinema* is still confined to home production, as there is no organized processing unit or factory of *kinema* production (Tamang 2000a). *Kinema*-making technology has not been recognized as a small-scale industry worthy of financial support or loan by any public sector bank or financial institution, nor has it been incorporated in the rural development programs of the governments in Nepal, India, and Bhutan.

3.1.1.4   Microorganisms

Bacilli: *Bacillus subtilis* (dominant functional bacterium); lactic acid bacteria (LAB): *Enterococcus faecium*; yeasts: *Candida parapsilosis* and *Geotrichum candidum* (Tamang 1992; Sarkar et al. 1994).

3.1.2   Hawaijar

Hawaijar is a traditional fermented soybean alkaline food of Manipur. It is prepared from a local variety of small-seeded soybean grown in hilly terraces of Manipur. It is similar to *kinema* (Figure 3.5).

3.1.2.1   Indigenous knowledge of preparation

Small soybean seeds are selected, washed, and boiled in an open cooker for 2–3 h. Excess water is drained off, and the product is cooled to ≈40°C and then the whole soybean seeds are packed in a small bamboo basket with a lid. The basket is lined with fresh leaves of fig plant (*Ficus hispida*), locally called *assee heibong* in the Meitei language, or banana leaves. After the cooled soybean seeds have been placed inside the basket, the lid is closed loosely and the basket is kept near the kitchen or a warm place for natural fermentation for 3–5 days (Figure 3.6). The emission of a typical ammonia odor and the appearance of a sticky texture on the cooked soybean seeds are considered to be a sign of good-quality *hawaijar* by the Meitei. Shelf-life of *hawaijar* is maximum 7 days without refrigeration. Sometimes, it is sun dried for 2–3 days and stored for several weeks for future consumption.

In contrast to *kinema* production, the practice of cracking and addition of ash is not adopted by the Meitei women in *hawaijar* production. *Hawaijar* is produced by the Meitei women, and men support in the process.

Figure 3.5 *Hawaijar* of Manipur.

Soybean

↓

Wash; boil for 2–3 h

↓

Loosely pack in a bamboo basket lined with leaves

↓

Ferment (25–40°C, 2–3 d)

↓

HAWAIJAR

Figure 3.6 Indigenous method of preparation of *hawaijar* in Manipur.

3.1.2.2   Culinary practices and economy

A special curry called *chagempomba* is commonly prepared by the Meitei in Manipur and is eaten with steamed rice. *Hawaijar* is eaten directly or used as a condiment or mixed with vegetables to make curry in the Manipuri cuisine.

Hawaijar is commonly sold in local markets throughout Manipur by the Meitei women. It is packed loosely in fresh leaves of fig or banana and is sold at the Ima market of Imphal for Rs. 40–50 per kilogram.

Since *hawaijar* is one of the most popular ethnic food products of the Meitei, its demand is always high. Despite of its popularity, there is no organized food sector for mass-scale production of *hawaijar* in Manipur. Like *kinema*, the production of *hawaijar* has not been recognized as a small-scale industry worthy of financial support by any financial institution in Manipur. The product is still prepared at home, and many women are dependent upon the product for their livelihood.

3.1.2.3   Microorganisms

Bacilli: *Bacillus subtilis* (dominant functional bacterium), *B. licheniformis*, *B. cereus*, and nonbacilli bacteria such as *Staphylococcus aureus*, *S. sciuri*, and *Alkaligenes* spp. (Jeyaram et al. 2008a; Tamang et al. 2009).

3.1.3   Tungrymbai

Tungrymbai or *turangbai* is an ethnic fermented soybean food of Khasi and Garo in Meghalaya. It is similar to *kinema*.

3.1.3.1   Indigenous knowledge of preparation

Soybean seeds are collected, cleaned, washed, and soaked in water for about 4–6 h (Agrahar-Murugkar and Subbulakshmi 2006). The seed coat of soybean is normally removed before cooking by rubbing the soaked seeds gently. The soaked soybeans are cooked for about 1–2 h until all the water is absorbed. Cooked beans are allowed to cool and are packed with fresh leaves of *Clinogyne dichotoma* (locally called *lamet*) and are placed inside a bamboo basket and covered by a thick cloth. The covered basket is kept over the fireplace and fermented naturally for 3–5 days to get *tungrymbai* (Figure 3.7).

3.1.3.2   Culinary practices and economy

Tungrymbai is mashed and put into a container with water and boiled until the water evaporates while being stirred continuously. It is mixed with fried onion, garlic, ginger, chilli, ground black sesame (locally called *til*), and salt. A thick curry is made and is served as a side dish with steamed rice by the Khasi in Meghalaya. Pickle is also made from *tungrymbai*.

Khasi women commonly sell *tungrymbai* packed in fresh leaves of *lamet* or banana (Figure 3.8) at the vegetable markets of Shillong. The price of each packet is Rs. 10, weighing 200–300 g per packet.

3.1.3.3   Microorganisms

Bacilli: *Bacillus subtilis* (dominant functional bacterium) and other *Bacillus* spp. (Tamang et al. 2009).

Figure 3.7 Indigenous method of *tungrymbai* preparation in Meghalaya.

3.1.4 Aakhone

Aakhone, also called *axone,* is an ethnic fermented sticky soybean food of Sema Naga in Nagaland, similar to *kinema.*

3.1.4.1 Indigenous knowledge of preparation

The preparation is the same as for other fermented soybean foods of North East India. Soybean seeds are soaked and cooked, and beans are wrapped in fresh leaves of banana or *Phrynium pubinerve* Blume (family: Marantaceae) or *Macaranga indica* Wight (family: Euphorbiaceae) and kept above the fireplace to ferment for 5–7 days (Mao and Odyuo 2007). The shelf life of freshly fermented *aakhone* is, at maximum, a week. Fresh *aakhone* is molded and made into cakes and then dried above an earthen

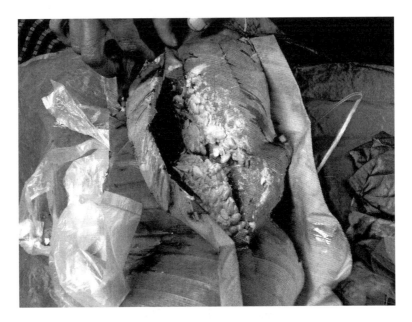

Figure 3.8 *Tungrymbai* of Meghalaya.

oven. Sometimes, each fermented bean is separated by hand and dried in the sun for 2–3 days. Such dried *aakhone* is stored in containers for future consumption (Figure 3.9).

3.1.4.2   Culinary practices and economy
Pickle is made from freshly fermented *aakhone* by mixing with green chilli, tomato, and salt. The dried *aakhone* cakes are cooked with pork and are eaten as a side dish with steamed rice by the Sema. Local women sell *aakhone* in several local vegetable markets in Nagaland.

3.1.4.3   Microorganisms
Bacilli: *Bacillus subtilis* (dominant functional bacterium) and other *Bacillus* spp. (Tamang et al. 2009).

3.1.5   Bekang

Bekang is an ethnic fermented soybean food commonly consumed by the Mizo in Mizoram. It is also similar to *kinema* (Figure 3.10).

3.1.5.1   Indigenous knowledge of preparation
During the traditional method of preparation of *bekang*, small, dry seeds of soybean are collected, cleaned, and soaked in water for 10–12 h. Excess water is drained off, and beans are boiled for 2–3 h in an open cooker

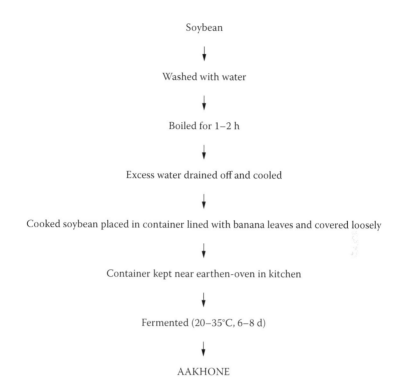

Soybean

↓

Washed with water

↓

Boiled for 1–2 h

↓

Excess water drained off and cooled

↓

Cooked soybean placed in container lined with banana leaves and covered loosely

↓

Container kept near earthen-oven in kitchen

↓

Fermented (20–35°C, 6–8 d)

↓

AAKHONE

Figure 3.9 Indigenous method of preparation of *aakhone* in Nagaland.

until the beans become soft. Excess water is drained off, and the beans are wrapped in fresh leaves of *Calliparpa aroria* (family: Verbanaceae), locally called *nuhlhan*, or leaves of *Phrynium* sp. (Family: Merantaceae), locally known as *hnahthial*. The wrapped beans are kept inside the small bamboo basket, which is then placed near an earthen oven or warm place and allowed to ferment naturally for 3–4 days. A sticky soybean product with an ammonia odor is produced, which is liked by the local consumers. The product is called *bekang* in Mizoram (Figure 3.11).

3.1.5.2   Culinary practices and economy
Bekang is consumed as it is or made into a curry with the addition of salt, green chillies, and tomatoes. It is consumed as a side dish with steamed rice. *Bekang* is sold in the local markets by Mizo women, who earn their livelihood by selling this product.

3.1.5.3   Microorganisms
Bacilli: *Bacillus subtilis* (dominant functional bacterium) and other *Bacillus* spp. (Tamang et al. 2009).

Figure 3.10 *Bekang* of Mizoram.

Soybean

↓

Soak overnight

↓

Boil for 1–2 h

↓

Dewater, cool

↓

Wrap in leaves and place inside basket

↓

Keep basket in warm place

↓

Ferment (20–35°C, 3–4 d)

↓

BEKANG

Figure 3.11 Indigenous method of preparation of *bekang* in Mizoram.

3.1.6 Peruyyan

Peruyyan is an ethnic fermented soybean food of the Apatani tribes in Arunachal Pradesh. The word *peruyyan* has been derived from the Apatani dialect; *perun* means beans, and *yannii* means packing in leaves.

3.1.6.1 Indigenous knowledge of preparation

During the traditional preparation of *peruyyan*, soybean seeds are collected, washed, and cooked for 2–3 h until the beans become soft. The excess water is drained off, and the beans are cooled for some time. The cooked soybeans are kept in a bamboo basket (vessel) lined with fresh ginger leaves, locally called *taki yannii*. The basket is loosely covered with ginger leaves and kept on a wooden rack above the fireplace for fermentation for 3–5 days. The stickiness of the product is checked, and if the product is sticky enough, then it is ready for consumption (Figure 3.12).

3.1.6.2 Culinary practices and economy

Peruyyan is consumed mostly as a side dish with steamed rice by the Apatani tribes in Arunachal Pradesh. It is mixed with hot water, chillies (locally called *tero*), and salt and directly consumed without frying or cooking, unlike *kinema* curry preparation.

3.1.6.3 Microorganisms

Bacilli: *Bacillus subtilis* (dominant functional bacterium) and other *Bacillus* spp. (Tamang et al. 2009).

3.2 Fermented black gram food

3.2.1 Maseura

Maseura or *masyaura* is an ethnic fermented black gram product of Darjeeling hills, Sikkim, and Nepal (Tamang 2000b). It is a cone-shaped, hollow, brittle, and friable product used as a condiment or an adjunct in cooking vegetables. *Maseura* is commonly prepared from split black gram or green gram or taro (*Colocasia*) tuber in Nepal (Karki 1994). It is one of the lesser-known fermented legume foods and is confined to the Newar castes of the Nepali (Figure 3.13).

Aerobic mesophilic counts in *maseura* are 10^8 cfu/g, and the loads of LAB and yeast are 10^7 cfu/g and 10^4 cfu/g, respectively (Chettri and Tamang 2008). The presence of a high number of LAB in *maseura* fermentation may be due to the predominance of LAB in dehulled black grams, which has been reported earlier in a similar black gram fermented product *idli* (Mukherjee et al. 1965). *Maseura* is slightly acidic, with the pH ranging

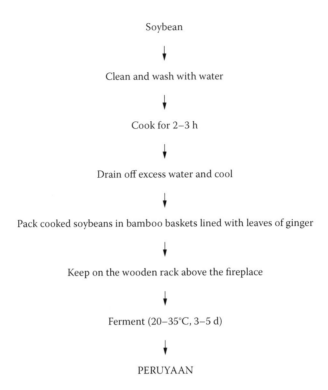

Soybean

↓

Clean and wash with water

↓

Cook for 2–3 h

↓

Drain off excess water and cool

↓

Pack cooked soybeans in bamboo baskets lined with leaves of ginger

↓

Keep on the wooden rack above the fireplace

↓

Ferment (20–35°C, 3–5 d)

↓

PERUYAAN

Figure 3.12 Indigenous method of preparation of *peruyyan* in Arunachal Pradesh.

from 5.6 to 6.3 (Chettri and Tamang 2008). Microorganisms isolated from *maseura* of Sikkim are bacteria (*Lb. fermentum, Lb. salivarius, P. pentosaceous, E. durans, Bacillus subtilis, B. mycoides, B. pumilus,* and *B. laterosporous*) and yeasts (*Saccharomyces cerevisiae, Pichia burtonii,* and *Candida castellii*), and molds are absent (Chettri and Tamang 2008). Microorganisms isolated from *maseura* of Nepal consist of bacteria (*P. pentosaceous, P. acidilactic,* and *Lb.* spp.), yeasts (*Saccharomyces cerevisiae* and *Candida versatilis*), and molds (*Cladosporium* spp., *Penicillium* spp., and *Aspergillus niger*) (Dahal et al. 2003).

3.2.1.1   Indigenous knowledge of preparation
During traditional preparation of *maseura* in Sikkim, seeds of black gram (*Phaseolus mungo* Roxb.) or green gram (*Phaseolus aureus* Roxb.) are cleaned, washed, and soaked overnight. Soaked seeds are split by pressing with the hands and the hulls are sloughed off. The split seeds are then ground into a thick paste using mortar and pestle. Water is carefully added while grinding until the paste becomes sticky, which is then

Figure 3.13 *Maseura* of Sikkim.

hand-molded into small balls or cones. If rice bean is used, then boiled potato or squash or yam is mixed with the paste to make it sticky. The mixture is placed on a bamboo mat and fermented in an open kitchen for 2–3 days, and then sun dried for 3–5 days, depending upon the weather conditions (Figure 3.14). *Maseura* can be stored in a dry container at room temperature for a year or more (Chettri and Tamang 2008).

The traditional method of preparation of *maseura or maysura* in Nepal is described as follows (Karki 1986; Dahal et al 2003). Black gram is split in a *janto*, a traditional stone grinder used for splitting and grinding dry grains. Cleaned, split black gram is soaked in water overnight, and the husk is removed by hand washing two to three times in water. The wet, dehulled black gram is then ground in a *silauto*, a grinder for wet grains. Spices or wet grain are placed over the stone slab and rubbed manually with the help of the stone ball. A small quantity of water is added to aid grinding to get a thick paste. *Colocasia* tuber is washed, peeled, shredded, and mixed with black gram *dal* paste in an approximate ratio of 1:1. The dough is then made into small lumps weighing 20–30 g each and distributed 1 to 2 in. apart on a bamboo tray. Usually, this operation is carried out in the evening, and the trays are then left overnight to ferment at room

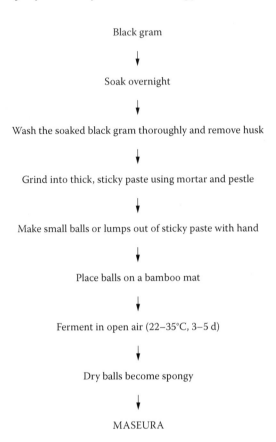

Figure 3.14 Indigenous method of preparation of *maseura* in Nepal and Sikkim.

temperature. The spongy, textured balls are then sun dried for 3–5 days to get *maseura*. It can be stored for several months at room temperature. *Maseura* is similar to North Indian *wari* or *dal bodi* and South Indian *sandige* (Dahal et al. 2005; Soni and Sandhu 1990a).

3.2.1.2 Culinary practices and economy
Maseura is used as a condiment or as an adjunct to vegetables in the Himalayas. It is usually fried in edible oil with vegetables to make a curry or soup and served with rice. In Nepal, it is used as a substitute for meat.

Production of *maseura* has drastically declined in many parts of Sikkim and Darjeeling hills (Chettri and Tamang 2008). The new generation hardly knows the product. The exact reasons have yet to be sorted out, but the unique indigenous method of *maseura* production is worth documenting as a low-cost fermented legume condiment in rural areas and also as a microbial resource in the Himalayas.

3.2.1.3 Microorganisms

Bacilli: *Bacillus subtilis, B. mycoides, B. pumilus, B. laterosporus*; lactic acid bacteria (LAB): *Pediococcus acidilactici, P. pentosaceus, Enterococcus durans, Lactobacillus fermentum, Lb. salivarius*; yeasts: *Saccharomyces cerevisiae, Pichia burtonii*, and *Candida castellii* (Chettri and Tamang 2008).

3.3 Microbiology

3.3.1 Kinema

3.3.1.1 Microorganisms

The population of bacilli in *kinema* is 10^8 cfu/g, followed by LAB at 10^7 cfu/g and yeasts at 10^4 cfu/g (Tamang 1992). Following the taxonomic keys of Norris et al. (1981), the endospore-forming rods are assignable to *Bacillus subtilis*, since they can produce catalase, amylase, and acetoin, but they are unable to grow anaerobically or to change the pH of Voges-Proskauer (VP) broth (Gordon et al. 1973) to less than 6.0. However, they do not share the characteristic of nitrate reduction of *B. subtilis*. Next to *B. subtilis*, the isolates have a close resemblance to *B. pumilus* and *B. coagulans*. The isolates differ from *B. pumilus* with respect to the position of spores, the hydrolysis of starch, and the ability to change the pH of VP broth to less than 6.0, and from *B. coagulans* with respect to growth in anaerobic agar or 7% NaCl and the ability to change the pH of VP broth to less than 6.0.

Following the detailed morphological and Analytical Profile Index (API) test profiles of *Bacillus* (Logan and Berkeley 1984), the bacilli show similarities mostly with *B. subtilis* and *B. coagulans*. They differ from *B. subtilis* with respect to production of tryptophan desaminase and acid from arbutin, salicin, and cellobiose. The presence of tryptophan desaminase and the central or paracentral position of spores in the *kinema* isolates make them different from *B. coagulans*. They differ from *B. pumilus* with respect to a number of characteristics, including position of spores, β-galactosidase and tryptophan desaminase content, and production of acid from D-galactose, N-acetylglucosamine, arbutin, salicin, cellobiose, insulin, starch, glycogen, and β-gentibiose (Sarkar et al. 1994).

According to the taxonomical criteria laid down by Claus and Berkeley (1986), *kinema* isolates are not unequivocally assignable to any of the species of *Bacillus*. They differ from *B. subtilis* with respect to position of spore, utilization of citrate, reduction of nitrate, and growth at 55°C. Inability to grow in anaerobic agar at 10°C and to change the pH of VP broth to less than 6.0, but the ability to hydrolyze gelatin and to grow in 5% and 7% NaCl without the use of growth factors differentiated the isolates from *B. coagulans*. Another closely related species, *B. pumilus*, differs from *B. subtilis* (Tamang 1992). Although the endospore-forming isolates have tentatively been designated as members of *B. subtilis*, they need the status of a new species.

The lactic acid bacterium is identified as *Enterococcus faecium*, and two types of yeasts are *Candida parapsilosis* and *Geotrichum candidum* (Tamang 1992; Sarkar et al. 1994). Besides *B. subtilis*, a number of species of *Bacillus* have been isolated from *kinema*, including *B. licheniformis*, *B. cereus*, *B. circulans*, *B. thuringiensis*, and *B. sphaericus* (Sarkar et al. 2002). However, *Bacillus subtilis* is the dominant functional bacterium in *kinema* (Sarkar and Tamang 1994; Tamang and Nikkuni 1996). *Bacillus subtilis* is an important starter culture for many Asian and African fermented soybean foods (Kiers et al. 2000). The traditional way of preparation of *kinema* and its use in curries is safe for consumption (Nout et al. 1998). There has been no record of outbreak of illness associated with *Bacillus subtilis* in fermented foods (Beumer 2001).

3.3.1.2   Source of inoculum in kinema production

Microbial analysis of raw soybeans collected from Darjeeling hills revealed the presence of endospores of *Bacillus subtilis* (Tamang 1992). Hesseltine (1983) earlier reported the presence of *Bacillus subtilis* on raw soybeans. Besides *B. subtilis*, populations of *E. faecium* and yeasts predominantly occurred during soaking of soybeans, indicating their entry through the water source (Tamang 2000c). *Enterococcus faecium* appears as a non-fecal contaminant (Mundt 1986). *Enterococcus faecium* has been reported in soaked soybeans during tempeh fermentation (Mulyowidarso et al. 1989). The load of enterobacteriaceae is below 10^4 cfu/g in overnight-soaked soybeans (Tamang 2003).

The use of a mortar and pestle during *kinema* production for cracking boiled soybeans also supplements the enterobacteriaceae population, which is detected at a level of 10^2 cfu/g. Microorganisms are killed during cooking of soaked soybeans except for heat-resistant spore formers. *Bacillus subtilis* is recovered in all sources, indicating its main role in fermentation (Tamang 2003). The source of *E. faecium*, *C. parapsilosis*, and *G. candidum* is mainly from the wooden mortar and pestle commonly used to crack cooked soybeans before wrapping during *kinema* production. Usually the wooden mortar and pestle are not washed properly by the rural *kinema* makers after cracking soybeans (Tamang 2002). This equipment is the main source of inocula (microorganisms) for spontaneous fermentation of *kinema*. Phylloplane studies of fresh fern fronds and fig leaves used as wrapping materials also suggest that these materials are supplemental sources of *B. subtilis* and *E. faecium* (Tamang 2003). Yeasts and molds are not recovered from these leaves, which are used as packing materials. Singh and Umabati Devi (1995) reported the presence of *Bacillus* and *Xanthomonas* spp. in fig leaves, which are used as packing material during *hawaijar* production.

It has been observed that the populations of yeasts as well as LAB increase remarkably in *kinema* kept at room temperature after desirable fermentation has been completed (Tamang 2003). *Bacillus* strains show a wide spectrum of enzymes, whereas strains of *Enterococcus* exhibit less enzymatic activity (Tamang 2003). *Bacillus* strains show strong peptidase and phosphatase activity but relatively weak esterase and lipase activity.

It has been observed that rich microbial diversity in various sources, particularly soybean, equipment, and leaves as wrapping materials, harness microbiota for spontaneous fermentation of *kinema* (Tamang 2000c). The practice of not cleaning up the mortar and pestle and using fresh leaves as wrapping materials by *kinema* producers reveals their indigenous knowledge of "microbiology" to preserve and supplement microorganisms for spontaneous fermentation of *kinema* without using starter cultures (Tamang 2003).

3.3.1.3   Optimization of fermentation period

The traditional methods of *kinema* preparation vary from home to home and, as a result, the quality of the product is inconsistent. To produce a product of reproducible quality with flavor and texture acceptable and attractive to larger groups, and to scale up production, the process parameters would need to be optimized by sensory evaluation.

The fermentation of beans either on a polyethylene sheet or within a tightly packed polyethylene bag gave a significantly unsatisfactory ($p < .05$) result. A thinly perforated polyethylene bag was determined to be the optimum packing material for fermenting soybean seeds. Although *Bacillus subtilis* does not grow anaerobically, the counts obtained under aerobic conditions and reduced oxygen conditions are the same (Tamang 1992). The colonies, however, spread better under reduced oxygen condition, producing larger colonies. The environment provided in the traditional process, as already described, and in the optimized perforated polyethylene bag would, in addition to providing a semi-anaerobic environment, also keep the soybeans moist, thus helping to spread the motile *B. subtilis* (Gordon et al. 1973) and facilitating the growth of the facultative anaerobic *E. faecium*.

It was further noted that soaked, but uncooked, soybeans produce a poor product, not only with respect to body and texture, but also regarding flavor and color. The cooking process reduces the undesirable microorganisms in situ (Wang and Hesseltine 1981). On the other hand, 25 min of cooking in 0.7-kg/cm^2 steam pressure seems to be high enough to kill even the important spore formers, thus resulting in significant deterioration ($p < .05$) of the product. The load of *B. subtilis* in raw soybeans as well as in the beans immediately after cooking was 10^6 cfu/g fresh weight.

The optimum time was identified as 10 min—enough to reduce the load of undesirable microorganisms developed during soaking without disturbing the load of heat-resistant endospore formers. Moreover, this temperature–time treatment provided the optimum softness of the seeds for fermentation. The temperature of incubation at 28°C resulted in a very slow fermentation rate and might provide the optimum milieu for growth of yeasts, making the product rancid. At 48-h fermentation time, 37°C was the optimum temperature for fermentation. The fermentation rate at 45°C was very high, and after 48 h of fermentation, the product was overfermented, resulting in significant deterioration ($p < .05$). Due to maintenance of high temperature for 48 h, the desirable viscous substance formed became dried. Traditionally, the fermentation time varies from 1 to 3 days. One day was not enough to complete fermentation at 37°C, perhaps because the growth of the fermenting microorganisms did not increase enough to cause the desirable biochemical changes (Tamang 1992). After 3 days of fermentation at 37°C, the product became dried-up and had a strong ammonia flavor and a rancid odor, resulting in a significantly low ($p < .05$) sensory score. Hence, the optimum fermentation time for production of *kinema* under optimized conditions was determined to be 48 h at 37°C (Sarkar and Tamang 1994).

3.3.1.4   In situ fermentation of kinema

In situ fermentation of *kinema* has been studied at 8-h intervals from 0 to 48 h, and the results showed that viable count of *B. subtilis* along with *E. faecium* and *C. parapsilosis* increases significantly in fermenting soybeans (Sarkar and Tamang 1995). Although acidity increased significantly at almost every 8-h interval, the fermentation was essentially alkaline, causing the pH of the beans to rise to about 8.5. With the decline in protein nitrogen content, the nonprotein and soluble nitrogen contents increased significantly at almost every 8-h interval during *kinema* fermentation (Tamang 1992).

The initial level of *B. subtilis*, even at the onset of fermentation, is due to its presence on raw soybeans (Hesseltine 1983) and passage through soaking and cooking treatments, the processes that do not reduce its load (Tamang 1992). Although *E. faecium* and *C. parapsilosis* are not found on raw soybeans, their detection even at the start of fermentation indicates their entry through tap water. In many foods, *E. faecium* appears as a nonfecal contaminant (Mundt 1986). *Candida parapsilosis* is present in water (Barnett et al. 1983). Unlike the 20–30-min cooking time in steam pressure for soybeans in *natto* preparation (Ohta 1986), the cooking time for soybeans in *kinema* preparation is only 10 min. In the latter process, the reason for the short cooking time is probably not to destroy the in situ endospore former, the main or possibly the sole fermenting organism, but to reduce the number of other organisms, because *kinema* is naturally fermented.

The number of *B. subtilis* cells increased significantly ($p < .05$) at 8-h intervals until the end of fermentation at 48 h. Although, initially the load of *E. faecium* is 40 times less than the load of *B. subtilis*, at the end of fermentation the load of *Enterococcus* is only 5 times less than that of *Bacillus* (Sarkar and Tamang 1995). This indicates that the growth rate of *Enterococcus* is even higher than that of *Bacillus,* even in alkaline pH. This is not surprising, because *E. faecium* is able to grow even at pH 9.6. Although present at a much lower load, *C. parapsilosis* increased significantly ($p < .05$) at 8-h intervals until 32 h.

During the first 16 h, as long as the organisms are growing exponentially, the pH of the fermenting beans go down from 6.9 to 6.6. During the first 8 h, there was no significant rise ($p < .05$) in free fatty acid and nonprotein nitrogen contents. Therefore, it seems likely that sugars, not proteins or fats, are initially used as substrates for metabolism and growth. Cooked soybeans contain sucrose, raffinose, and stachyose (Steinkraus 1983), and *B. subtilis* is capable of producing acid from sucrose and raffinose and their hydrolyzing products, glucose and fructose. *Enterococcus faecium* is capable of producing acid from galactose, another hydrolyzing product of raffinose. *Candida parapsilosis* is capable of utilizing sucrose and all the hydrolytic products of raffinose. However, after 16 h, the pH increases significantly ($p < .05$) at 8-h intervals until 40 h. This is probably due to the proteolytic activities of the microorganisms. While *B. subtilis* is capable of hydrolyzing protein (casein and gelatin), *E. faecium* is unable to hydrolyze them. Presumably, the protease produced by *B. subtilis* degraded soy proteins, which results in a significant increase ($p < .05$) of nonprotein nitrogen content at 8-h intervals, starting from 8 h until the end of fermentation. The production of ammonia is also common in *natto* (Hesseltine and Wang 1967; Kiuchi 2001). Due to the lipolytic activities of the microorganisms, the fat content in soybeans was significantly degraded ($p < .05$) to free fatty acids at 8-h intervals until the end of fermentation. A similar increase in free fatty acidity is also reported in *natto* (Hosoi and Kiuchi 2003).

3.3.1.5   *Selection of starter culture*

Sterile soybeans inoculated with only cells of *E. faecium* do not produce *kinema,* giving the lowest scores in every sensory attribute (Sarkar and Tamang 1994). Because *E. faecium* is unable to hydrolyze protein, soybeans are not fermented to *kinema* by this bacterium alone. Again, the product produced by any combination of yeast with *B. subtilis* had significantly lower ($p < .05$) scores than *kinema* produced by *B. subtilis* alone or in combination with *E. faecium.* This is due to the rancid flavor that develops during the fermentation. The sensory analysis shows that the yeasts *Candida* and *Geotrichum* have no role in fermentation of soybeans to make *kinema* (Sarkar and Tamang 1994). The combination of *B. subtilis* and *E. faecium* gives a satisfactory score, although it differs significantly ($p < .05$) from the product prepared solely by *B. subtilis. Kinema* prepared by a pure culture strain of *B. subtilis* scored the highest acceptability in sensory evaluation tests and was liked by all consumers (Sarkar and Tamang 1994). Based on these results, it is concluded that *B. subtilis* is the sole fermenting organism for the production of *kinema* (Tamang 1992; Sarkar and Tamang 1994).

The traditional method of *kinema* preparation results in a product with inconsistent quality. Pure strains of *B. subtilis*, previously isolated from market samples of *kinema*, were selected as possible starter cultures on the basis of enzyme activities and the production of slimy material (Tamang and Nikkuni 1996). Protease activity (U/ml), α-amylase activity (U/ml), and relative viscosity of the selected strains were 6.5–81.5, 0.1–9.3, and 1.1–20.1, respectively. *Kinema* produced by these strains showed nitrogen contents (expressed as a percentage of the total nitrogen content) of: water-soluble N, 48.4%–76.5%; trichloroacetic acid (TCA)-soluble N, 16.0%–27.6%; formal N, 5.0%–12.5%; and ammonia-N, 4.4%–7.8%. Reducing sugar increased up to 20% of the wet weight, and relative viscosity also increased from 2.0 to 35.2 (Tamang 1995).

Correlation matrices of the biochemical parameters and sensory attributes of the *kinema* produced by *B. subtilis* strains were statistically analyzed. There was a significant correlation at the 0.1% level between general acceptability and stickiness of *kinema*, and between general acceptability and taste of the product (Tamang and Nikkuni 1996). There was also a significant correlation at the 0.1% level between the taste and stickiness attributes of *kinema*. This datum helps to formulate the sensory quality of *kinema* by considering stickiness and taste as the most important attribute indices for general acceptability of *kinema* (Tamang 1995). The quality of *kinema* is mainly judged by its stickiness and flavor (Tamang 1992).

Bacillus subtilis, strains KK-2:B10 and GK-2:B10, were selected as the best starter cultures for monoculture fermentation of *kinema* from several native strains of *B. subtilis* previously isolated from traditionally fermented *kinema* (Tamang 1995; Tamang and Nikkuni 1996). These strains have been deposited at the Microbial Type Culture Collection (MTCC) of the Institute of Microbial Technology, Chandigarh, India: *Bacillus subtilis* KK-2:B10 (MTCC 2756) and *Bacillus subtilis* GK-2:B10 (MTCC 2757).

3.3.1.6   Monoculture fermentation of kinema

Kinema was prepared by using a monoculture strain of *B. subtilis* KK2:B10 under optimized conditions in the laboratory (Tamang and Nikkuni 1998). During monoculture fermentation, the growth rate of viable cells of *Bacillus subtilis* KK2:B10 was faster at 45°C until 16 h. The surface of the fermenting soybeans changed to a whitish color due to growth of *B. subtilis* on or in between the soybeans during fermentation. *Kinema* prepared at 40°C showed a remarkable increase in relative viscosity from 16 to 20 h compared with *kinema* prepared at 35°C and 45°C (Tamang and Nikkuni 1998). The unique feature of *B. subtilis* KK2:B10 is the formation of sticky viscous materials, which starts from 8 to 12 h at all incubation temperatures. Stickiness is an important criterion for judging quality of *kinema* by consumers (Tamang 1992).

Water-soluble nitrogen and formol nitrogen to total nitrogen contents of *kinema* increased rapidly during fermentation. This is due to the high proteolytic activity of *B. subtilis* KK2:B10 (Tamang 1995). Reducing sugars increased at log phase and then decreased sharply during *kinema* fermentation. This indicates that reducing sugars of fermenting soybeans are used by *B. subtilis* for its metabolism, since *B. subtilis* KK2:B10 produces a high activity of α-amylase (Tamang and Nikkuni 1996). Kanno et al. (1982) reported that glucose and fructose increased up to 8 h and then decreased rapidly during *natto* fermentation by *B. subtilis* (*natto*).

The effect of temperature on the maturation of freshly prepared *kinema* was studied, and the results showed a significant increase in relative viscosity of *kinema* during maturation at 5°C and 10°C (Tamang and Nikkuni 1998). However, no significant differences ($p < .05$) in water-soluble nitrogen and formol nitrogen content of *kinema* was observed during maturation at low temperatures compared with *kinema* kept at the control temperature. Keeping freshly prepared *kinema* below 10°C for 1 day stabilized the quality of the product by preventing the further biological activity of microorganisms and showed better stickiness, which is very important sensory property of *kinema* (Tamang 1995). This maturation process may improve the unique flavor of *kinema*. Maturation of *natto* below 10°C for 1–2 days after fermentation is one of the major steps in commercial *natto* production for flavor and viscosity development (Ueda 1989).

The quality of *kinema* prepared by starter culture of *B. subtilis* KK2:B10 (MTCC 2756) was maintained by using a 20-h fermentation period at 40°C followed by maturation at 5°C for 24 h. *Kinema* produced under these optimized conditions has more advantages over the traditional method due to shorter fermentation time, better hygienic conditions, greater consistency in quality, and increased levels of soluble protein (Tamang 1998b). Organoleptically, the monoculture fermentation of soybeans by *B. subtilis* produces the best *kinema* because of a pleasant nutty flavor and highly sticky texture of the product. The shorter time obtained for optimum fermentation eliminates the chance of growth of contaminants and minimizes high levels of ammonia, which adversely affects the nutritional quality and palatability of the product (Odunfa and Adewuyi 1985).

3.3.1.7 Development of pulverized starter for kinema production

Nutrient broth is conventionally used for enrichment of *Bacillus subtilis* spores. Nutrient broth is composed of expensive beef extract, which is not acceptable to the majority of the Hindu population in the Himalayas. Moreover the crude soybean extract after cooking soybeans is discarded during *kinema* preparation. Instead of discarding the soybean extract after autoclaving soybeans, inexpensive soybean extract broth was used as a medium for enrichment of *B. subtilis* spores (Tamang 1999). The load of *Bacillus subtilis* KK2:B10 was significantly (p <.05) higher in the soybean crude extract broth (3.2×10^8 cfu/ml) as compared with the nutrient broth (0.4×10^8 cfu/ml) and a phytone-sucrose broth (0.2×10^8 cfu/ml) (Tamang 1999). Hence, soybean crude extract, after adjusting pH to 7.0, was selected as an inexpensive broth medium for enriching spores of *B. subtilis* for monoculture fermentation of *kinema*.

A ready-to-use pulverized starter for *kinema* production was prepared (Figure 3.15). *Kinema* prepared by using *B. subtilis* KK2:B10 strain that was harvested in soybean extract broth was dried in an oven at 70°C for 10 h and ground aseptically to make pulverized starter. The 1% of pulverized starter instead of *B. subtilis* was added aseptically to autoclaved soybeans and fermented to produce *kinema* (Figure 2.8, Stage A). The total viable count of *B. subtilis* in pulverized starter was constantly maintained at the level of 10^9 cfu/g for 6 months. This is due to survival of endospores of *B. subtilis* for a longer period at room temperature. No other microorganisms were recovered from the pulverized starter, which was kept in a sterile polyethylene bag at room temperature (Tamang 1999).

The consumers' preference trials showed that *kinema* prepared by using pulverized starter under optimized conditions was more acceptable than market *kinema* (Tamang 1999). Water-soluble nitrogen and formol nitrogen contents were higher in *kinema* prepared by using pulverized starter than market *kinema* (Tamang 1999). Increased water-soluble nitrogen in *kinema* helps in digestibility, and a high amount of formol nitrogen contains free amino acids supplements that impart better taste to *kinema* (Nikkuni et al. 1995).

Application of ready-to-use pulverized starter may appear appropriate in *kinema* production for marginal *kinema* producers in the eastern Himalayas, since it is cost effective and easy to handle. *Kinema* prepared using pulverized starter has several advantages over the traditional method, including a shorter fermentation time, and greater product consistency with better quality, maximum stickiness, and less ammonia odor.

3.3.1.8   *Phylogenetic similarity of Bacillus strains from Asian fermented soybeans*

The phylogenetic relationship among bacilli isolated from *kinema* (India), *chungkokjang* (Korea), and *natto* (Japan), and similar fermented sticky soybean foods of Asia, has been studied on the basis of 16S rDNA sequence (Tamang et al. 2002). Strains of *Bacillus* isolated from *kinema* and *chungkokjang* showed central to paracentral position of spores, with a few strains showing a negative nitrate reduction test, whereas *Bacillus subtilis* (*natto*) strains isolated from *natto* showed a central position of spores, and all

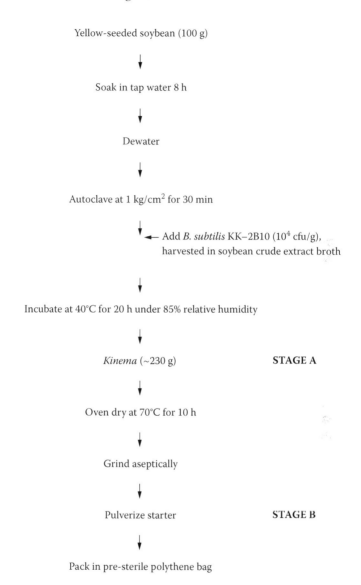

Figure 3.15 Flow sheet of pulverized *kinema* starter production.

reduced nitrate (Tamang et al. 2002). However, all strains of *Bacillus sub-tilis* isolated from *kinema*, *chungkokjang*, and *natto* showed stickiness on phytone agar and cooked soybean, which are characteristic properties of nonsalty fermented soybean foods of Asia (Tamang and Nikkuni 1996). However, the *B. subtilis* JCM 1465 strain did not produce any stickiness (Tamang et al. 2002).

In order to investigate the phylogenetic relationship of isolates to other bacteria, the sequence of 16S rRNA gene PCR product was determined, and the results showed that *B. subtilis* strains KD:B1 and KG:B1 isolated from *kinema*, *B. subtilis* CA:B1 and CK:B2 isolated from *chungkokjang*, and *B. subtilis* JN-1 isolated from *natto* have identical sequences except that of JA-1 (*natto*), which has one ambiguous nucleotide (Tamang et al. 2002). The evolutionary distance between four strains—CK:B1, KD:B1, JN-1, and JA-1—and *Bacillus subtilis* was 0.002 K_{nuc} as calculated by the ratio of nucleotide substitution per nucleotide site, indicating 99% homology with *Bacillus subtilis* type strain. However, the evolutionary distance between the strains CA:B1 and KG:B1 and *Bacillus subtilis* was 0.005 K_{nuc}, showing approximately 99.5% homology with *Bacillus subtilis* type strain (Tamang et al. 2002). The phylogenetic analyses revealed that all six strains belonged to *B. subtilis*. This is the first report to describe the phylogeny of *B. subtilis* isolated from similar nonsalty fermented sticky soybean foods of Asia (Tamang et al. 2002). The plasmid of *B. subtilis* (*natto*) strain resembles that of *B. subtilis* strain in the partial nucleotide sequences (Hara et al. 1995). The diversity of *Bacillus subtilis*-fermented soybean foods of Asia, including that of the eastern Himalayas, needs to be studied in detail to identify their similarities. The probable source of common stock of similar sticky fermented soybean foods will help food scientists to trace the antiquity of fermented soybean foods in Asia.

Although *E. faecium* does not add any sensory quality to the *Bacillus* fermentation of soybeans, it is always encountered in naturally fermented *kinema* (Tamang 1992). The presence and growth of yeast during *kinema* preparation are associated with the development of rancidity in the products. In fact, *B. subtilis* is the sole fermenting organism in *kinema* preparation.

3.3.2   Other fermented soybean foods of North East India

Tungrymbai, bekang, aakhone, and *peruyyan* have been microbiologically investigated, and *hawaijar* has been preliminarily studied. *Bacillus subtilis, B. licheniformis, B. cereus, Staphylococcus aureus, S. sciuri, Alkaligenes* spp., and *Providencia rettgeri* have been isolated from samples of *hawaijar* of Manipur (Jeyaram et al. 2008a). A genetic diversity of *B. subtilis* phylogenic groups was investigated, and it was found that *B. subtilis* was the dominant bacterium in *hawaijar*, as in *kinema* (Jeyaram et al. 2008a). *Bacillus subtilis* has been isolated from *tungrymbai, aakhone, hawaijar, peruyyan,* and *bekang* samples (Tamang et al. 2009).

3.4   Nutritive value

Kinema has a high ($\approx$62%) moisture content because of the soaking and cooking of soybeans prior to fermentation (Table 3.1). A marked decrease

Table 3.1 Nutritional Composition of *Kinema*

Parameters	Raw soybean	*Kinema*
Moisture (%)	10.8	62.0
Ash (100 g/DM)	5.0	7.2
pH	6.7	7.9
Protein (100 g/DM)	47.1	47.7
Fat (100 g/DM)	22.1	17.0
Carbohydrate (100 g/DM)	25.8	28.1
Food value (kcal/100 g DM)	478	454
Fe (mg/100 g)	8.7	17.7
Mn (mg/100 g)	2.7	5.4
Zn (mg/100 g)	3.8	4.5
Na (mg/100 g)	1.7	27.7
Ca (mg/100 g)	186.0	432.0
Total amino acids (mg/100 g)	43,654	46,218
Free amino acids (mg/100 g)	472	5,129

Note: DM, dry matter.

in the fat content of *kinema* compared to raw soybeans is due to the lipolytic activities of the microorganisms during *kinema* production, with a concomitant increase in free fatty acidity (Tamang 1992). The higher ash content of market samples of *kinema* is likely due to addition of firewood ash during production. Although the acidity in *kinema* increases by about onefold over raw soybeans, the product has a high pH value of 7.9. This is due to the high buffering capacity of the legume beans and the proteolytic characteristic of most vegetable protein fermentations (Hesseltine 1965). The food value of *kinema* is 454 kcal/100 g dry matter (Tamang 1992), which is almost the same as other Asian fermented soybean foods (Campbell-Platt 1987).

Based on the cost of protein per kilogram, *kinema* is the cheapest source of plant protein, as compared to milk and animal products, that is easily accessible to the rural poor of the Himalayas. During the process of *kinema* production, soy proteins, which have been denatured by cooking, are hydrolyzed by proteolytic enzymes produced by *Bacillus subtilis* into peptides and amino acids that enhance digestibility (Tamang and Nikkuni 1998). A remarkable increase in water-soluble nitrogen and trichloroacetic acid (TCA)-soluble nitrogen contents have been observed during *kinema* fermentation (Sarkar and Tamang 1995). Total amino acids, free amino acids, and mineral contents increase during *kinema* fermentation (Table 3.1), and subsequently enrich the nutritional value of the product (Sarkar and Tamang 1995; Nikkuni et al. 1995; Sarkar et al. 1997a; Tamang and Nikkuni 1998). *Kinema* contains all essential amino acids, and its essential amino acids score is as high as that of egg and milk proteins (Sarkar et al. 1997a). Degradation of oligosaccharides has been reported in *kinema* (Sarkar et al. 1997b). *Kinema* is rich in linoleic acid, an essential fatty acid in foods (Sarkar et al. 1996). Phytosterols, which have a cholesterol-lowering effect, are increased during *kinema* fermentation (Sarkar et al. 1996). Traditionally prepared *kinema* contains 8 mg thiamine, 12 mg riboflavin, and 45 mg niacin per kg dry matter (Sarkar et al. 1998). The content of riboflavin and niacin increases in *kinema*, while thiamine decreases during fermentation (Sarkar et al. 1998). An increase in total phenol content from 0.42 mg GAE (gallic acid equivalent)/g in boiled soybean to 2.3 mg GAE/g in *kinema* has been observed (Tamang et al. 2009). Total phenol content in food is one of the indicators of antioxidant activity (Pourmorad et al. 2006). An increase in total phenol content has also been reported in the Korean fermented soybean food *chungkokjang* (Shon et al. 2007) and the Chinese fermented soybean food *douchi* (Wang et al. 2007). The flavor of *kinema* originates from the hydrolysis of soybean proteins to peptides and amino acids, fermented by a pure culture of *B. subtilis* (Tamang 1992). *Kinema* has many health-promoting benefits, including antioxidant properties, digested protein, essential amino acids, vitamin B complex, low-cholesterol content, etc.

Compared with nonfermented soybean flour, *kinema* flour has a higher viscosity and nitrogen solubility index (Shrestha and Noomhorn 2001). The recipe of *kinema* has been diversified and made into biscuits containing higher protein and improved organoleptic property (Shrestha and Noomhorn 2002). *Kinema* is a highly nutritious food containing 48 g protein per 100 g dry matter (Tamang 1992). Tungrymbai contains protein (45.9 g/100 g), fat (30.2 g/100 g), fiber (12.8 g/100 g), carotene (212.7 µg/100 g), and folic acid (200 µg/100 g) (Agrahar-Murugkar and Subbulaksmi 2006).

Soaking of raw soybean seeds and then discarding the water and cooking are the important steps in preparation of *kinema* and similar products. Soaking and cooking of soybeans help to inactivate and leach out some undesirable factors such as phytic acid (Chang et al. 1977), flatus-causing oligosaccharides (Wang et al. 1979), and trypsin inhibitors (Albrecht et al. 1966), the water of which is then discarded (Hesseltine 1985a). The heating process also reduces the in situ microbial contaminants, especially nonsporulating bacteria and molds (Wang and Hesseltine 1981).

The ammonia odor with typical *kinema* flavor is always acceptable to *kinema* consumers. However, some market samples analyzed had an undesirable rancid odor. This may be due to the liberation of free fatty acids in higher amounts by lipolytic activities of microflora present in *kinema* (Tamang 1992). Yong and Wood (1977) observed that the liberation of free fatty acid in higher amounts is undesirable because of its own taste and its contribution to developing rancidity in the product. Kiuchi et al. (1976) observed that a high amount of free fatty acid in *hama-natto* gives the product a strong, harsh taste. However, Wang et al. (1975) identified the free fatty acids liberated during fermentation of soybeans by *Rhizopus oligosporus* as antitryptic factors. The *kinema* production is a solid state fermentation process. The high moisture content, however, indicates its very short shelf life. To improve its keeping quality, the local people often sun dry *kinema*. Although no toxicological study has been conducted with *kinema*, information regarding the toxic effect due to consumption of *kinema* has not been reported. The fact that molds are not found in this fermentation makes *kinema* safe from the risk of mycotoxins. *Bacillus subtilis* is nonpathogenic and therefore a safe organism (Odunfa 1981).

Maseura is slightly acidic, with the pH ranging from 5.6 to 6.3 (Table 3.1). Moisture content of dried *maseura* is about 8%–10%; protein is 18%–20%; carbohydrate is 67%–70%; and there is a trace of minerals (Dahal et al. 2003). Increases in soluble protein, amino nitrogen, nonprotein nitrogen, thiamine, and riboflavin have been observed in *maseura* of Nepal (Dahal et al. 2003). The essential amino acid profile does not show any significant change, while sulfur amino acids are found to be the first limiting amino acids in *maseura* (Dahal et al. 2003).

3.5   Conclusion

The flavorsome and mucilaginous fermented soybean foods *kinema*, *hawaijar, tungrymbai, bekang, aakhone*, and *peruyyan* are popular among the Mongolian-origin races in the eastern Himalayas. The Mongolian people prefer the umami-flavored foods due to specific sensory development. *Bacillus subtilis* is the dominant functional bacterium in all fermented soybean foods of these regions.

Fermented soybean foods are consumed only in the eastern Himalayas; no such product is consumed in other parts of the Himalayas. Although there is a good demand for ethnic fermented soybean foods among the local consumers in North East India, the production is limited to the household level. In the interest of generating income, ready-to-use pulverized starter culture for *kinema* production could be introduced to makers of *kinema* or similar sticky fermented soybean foods of North East India adapted to local conditions.

Scientific findings have corroborated the indigenous knowledge of the ethnic people of the Himalayas and acknowledged the innovative skills of mountain women in producing fermented soybeans. Ethnic fermented soybeans are one of the major food resources in the eastern Himalayas. They provide an inexpensive, highly digestible plant protein with low fat/cholesterol content, high nutritive value, and antioxidant and other health-promoting properties in the local diet as a functional food.

chapter four

Fermented milks

Milk and milk products are the major food components in the dietary culture of ethnic people living in the different elevations and climatic zones of the Himalayas, where agrarian and pastoral systems coexist. Cow milk is commonly consumed throughout the Himalayan regions, and buffalo milk is more popular in the foothills of the western Himalayas. Yak (*chauri* in Nepali and *kno*, the female yak, in the Tibetan language) is commonly used in subalpine and alpine Himalayas as a source for milk, meat, skin, and hair in the highlands above elevation 2100 m. Consumption of goat milk is uncommon in the Himalayas.

4.1 Important fermented milk products

The Himalayan people prepare and consume a variety of ethnic fermented milk products made from cow milk and yak milk, such as such as *dahi*, *mohi*, *gheu*, *soft chhurpi*, *chhu*, *somar*, *maa*, *philu*, and *shyow*. Some of the milk products, notably *dahi* (curd), *mohi* (buttermilk), and *gheu* (butter), are familiar in all regions of the Himalayas; *chhurpi*, *chhu*, *philu*, etc., are confined mostly to the Tibetans. *Somar* is exclusively prepared and consumed by the Sherpa of Nepal and Sikkim living at high altitude. The practice of standard starter culture is uncommon; however, the back-sloping technique, i.e., the use of previously fermented product as a source of microorganisms(s) into the freshly boiled milk, is a common practice. Milk and milk products are not the traditional food items in the food culture of many ethnic people of North East India, where the pastoral system is not a traditional agriculture or livestock system. Rearing of animals is mostly for meat. The Aryan-origin ethnic people who live in the sub-Himalayan regions prefer mild-flavored milk products, whereas the Mongolian-origin ethnic people living at high altitude prefer to have strong-flavored milk products.

4.1.1 Dahi

Dahi (curd) is a popular fermented milk product of the Himalayas and is also used for the preparation of a number of other ethnic milk products: *gheu*, *mohi*, soft *chhurpi*, *chhu*, etc. *Dahi* is the Nepali and Hindi word; the Tibetans call it *shyow*.

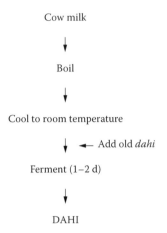

Cow milk

↓

Boil

↓

Cool to room temperature

↓ ◄— Add old *dahi*

Ferment (1–2 d)

↓

DAHI

Figure 4.1 Indigenous method of *dahi* preparation in the Himalayas.

4.1.1.1   Indigenous knowledge of preparation

Fresh milk of cow or yak is boiled in a vessel, cooled to room temperature, and transferred to a hollow wooden vessel or container locally
known as a *theki*. A small quantity of previously prepared *dahi* (called
mau in Nepali) serves as a source of inoculum and is added to boiled and
cooled milk. This is left for fermentation for 1–2 days in summer and for
2–4 days in winter at room temperature (see Figure 4.1). The duration of
fermentation depends on the season as well as on the geographical location of the place.

4.1.1.2   Culinary practices and economy

Dahi is consumed directly as a refreshing nonalcoholic beverage and
savory. Sugar may or may not be added before consumption. It is also
consumed after mixing it with rice or *chiura* (beaten rice). *Dahi*, being very
popular, finds a good market all over the Himalayas. The price of *dahi*
ranges from Rs. 16 to Rs. 25 per kilogram in the local markets.

4.1.1.3   Microorganisms

Lactic acid bacteria (LAB): *Lactobacillus bifermentans*, *Lb. alimentarius*, *Lb.
paracasei* subsp. *pseudoplantarum*, *Lactococcus lactis* subsp. *lactis*, and *Lc. lactis* subsp. *cremoris*; yeasts: *Saccharomycopsis* spp. and *Candida* spp. (Dewan
and Tamang 2007).

4.1.2   Mohi

Mohi or buttermilk is a liquid fermented milk product. It is usually a by-
product of *dahi*. *Mohi* is the Nepali term meaning "buttermilk." The Bhutia

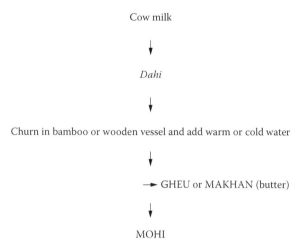

Cow milk

↓

Dahi

↓

Churn in bamboo or wooden vessel and add warm or cold water

↓

→ GHEU or MAKHAN (butter)

↓

MOHI

Figure 4.2 Indigenous method of *mohi* and *gheu* preparation in the Himalayas.

and Lepcha call it *kachhu*. In the western Himalayas, buttermilk is known as *lassi*, which is a popular refreshing beverage.

4.1.2.1   Indigenous knowledge of preparation
Dahi is churned to produce *gheu* (butter), and the liquid that remains is *mohi* (Figure 4.2).

4.1.2.2   Culinary practices and economy
Mohi is consumed as a cooling beverage during hot days and also to overcome tiredness. It is also processed further to produce other fermented milk products like soft *chhurpi*, *chhu*, *dudh chhurpi*, etc. *Lassi* is drunk with addition of sugar in summer as a refreshing beverage in Uttarakhand, Himachal Pradesh, and Jammu & Kashmir.

4.1.2.3   Microorganisms
Lactic acid bacteria (LAB): *Lactobacillus alimentarius*, *Lactococcus lactis* subsp. *lactis*, and *Lc. lactis* subsp. *cremoris*; yeasts: *Saccharomycopsis* spp. and *Candida* spp. (Dewan and Tamang 2007).

4.1.3   Gheu

Gheu or *ghee* is a locally prepared butter from cow milk, with a typical flavor and aroma, and is produced by churning *dahi*. It is a common milk product of the Himalayas and is consumed by the entire population. It is important commercially as well as for various other functions. *Gheu* is a Nepali word for *ghee* or *makhan* in Hindi, *maa* in Tibetan, and *mor* in Lepcha.

4.1.3.1   Indigenous knowledge of preparation

Dahi is churned to produce *gheu* (butter) and *mohi* (buttermilk). The churning can be done in a number of ways (Figure 4.2). The most common and easy method is by using a long bamboo vessel, variously called *tolung*, *somg*, or *padung*. *Madani* is made of a long stick (*shar*) with a circular or star-shaped flat wooden disc (*pangra*) at one end. *Dahi* is poured into the bamboo vessel and is churned by lifting and lowering of the *madani* inside the bamboo vessel. The churning is done for 15–30 min, with the addition of either cold or warm water, as the weather demands, to facilitate better separation of the *gheu* from the liquid *mohi*. In the process of this churning, a big lump of soft *gheu* is formed and seen floating on the *mohi*. It is carefully lifted out with the hand and transferred to another vessel. Another method of churning is by using a *theki*, a hollow wooden vessel, and a *madani*, consisting of the *ghurra* and a string called a *neti*. *Dahi* is kept inside the *theki* and churned by pulling the *neti* with either hand so that *madani* rotates in alternating clockwise and anticlockwise directions in *theki*. *Gheu* is collected as in the first method.

In North Sikkim, Ladakh, some parts of Arunachal Pradesh, North and West Bhutan, and Tibet, butter is prepared from yak milk called *maa*, which is stored by wrapping it in a piece of cube-shaped dried skin of sheep and by stitching all the edges from inside. This type of butter, known as *maata*, can be stored for several months or even years (Figure 4.3).

Figure 4.3 *Maa* (yak butter) stored in sheep skin called *maata* in Sikkim.

4.1.3.2   Culinary practices and economy

Freshly prepared *gheu* (or *nauni gheu*, as locally called by the Nepali) is also consumed as it is. It is purified further by boiling until the oily liquid separates from the unwanted dark-brown precipitate, locally called *khar*. The purified *gheu* is then consumed along with steamed rice or mixed in *dal*, curry, etc. It is also spread over chapati or baked bread, mostly in the western Himalayas. *Gheu* is used to prepare traditional cereal-based snacks and varieties of sweets. *Maa* is used for cooking and frying edible items, or it is consumed directly by melting with barley or baked potato by the Tibetans.

Gheu is a highly prized milk product and serves as a major source of income for farmers in the Himalayas. It is sold in the local markets all the year round. It costs about Rs. 130–200 per kilogram.

4.1.3.3   Microorganisms

Lactic acid bacteria (LAB): *Lactococcus lactis* subsp. *lactis* and *Lc. lactis* subsp. *cremoris* (Dewan 2002).

4.1.4   Chhurpi (hard variety) and dudh chhurpi

Hard-variety *chhurpi* is mostly prepared from yak milk at high altitudes (2100–4500 m) and has characteristic gumminess and chewiness, and it is commonly eaten as a masticator (Katiyar et al. 1991; Pal et al. 1996). Two types of hard-*chhurpi* are prepared: common *chhurpi* and *dudh chhurpi* (Hossain et al. 1996; Pal et al. 1993; Tamang 2000b). *Chhurpi* is also called *khamu* by the Tibetans (Figure 4.4).

4.1.4.1   Indigenous knowledge of preparation

Cream is separated from milk, and the skimmed milk is boiled and curdled by adding whey. After straining, the coagulum is cooked until the remaining water dries up. The highly stringy mass is wrapped in a cloth and fermented under pressure at room temperature for about 2 days. After pressing, the mass is sliced and allowed to dry by placing it above an earthen oven for about a month (Figure 4.5).

During the preparation of *dudh chhurpi*, curd is churned in a bamboo vessel for an hour to produce butter, which is separated from the buttermilk. Buttermilk (*mohi*) is cooked in a vessel over a fire for 1–2 h to make a solid white mass. This is taken out using a bamboo sieve, the liquid is drained off, and the mass is placed inside a sack. This mass is pressed with a heavy stone for 2–4 h to drain out the liquid further. The solid is taken out of the sack and cut into cubes weighing about 6 g. *Thake*, a thick

Figure 4.4 Hard varieties and *dudh chhurpi* of Bhutan.

paste prepared from the cooked whey and skim milk cream, is applied on the surface of the pieces, which are then threaded and hung in the open or above a kitchen oven for drying for 7–10 days (Figure 4.6). Sweet and dried skim-milk powder covers the hard texture of the *chhurpi*; hence it is called *dudh* (milk in local language) *chhurpi*.

4.1.4.2 Culinary practices and economy

Dudh chhurpi is available as small cube-shaped solids. Both varieties of hard-*chhurpi* are consumed as a nutritious masticator like a chewing gum in all Tibetan areas in the Himalayas. *Chhurpi* chewing gives extra energy at high altitudes.

In the market, *dudh chhurpi* pieces are hung in long threads. The hard-variety *chhurpi* is sold in all local periodical markets, mostly by women. One kg of *chhurpi* costs about Rs. 150 or more. The cost analysis of *dudh chhurpi* production in Darjeeling hills, Sikkim, and Bhutan shows a net annual profit of 32% on the total investment (Hossain et al. 1996). *Chhurpi* production is a promising small-scale enterprise in the Himalayas.

4.1.4.3 Microorganisms

Lactic acid bacteria (LAB): *Lactobacillus farciminis, Lb. paracasei* subsp. *paracasei, Weisella confuses,* and *Lb. bifermentans* (Dewan 2002).

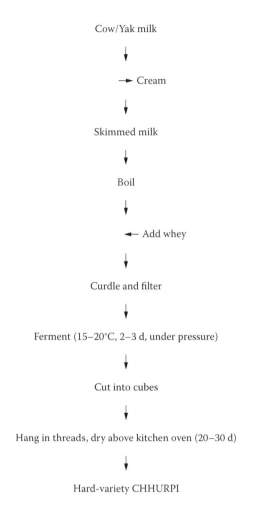

Cow/Yak milk

↓

→ Cream

↓

Skimmed milk

↓

Boil

↓

← Add whey

↓

Curdle and filter

↓

Ferment (15–20°C, 2–3 d, under pressure)

↓

Cut into cubes

↓

Hang in threads, dry above kitchen oven (20–30 d)

↓

Hard-variety CHHURPI

Figure 4.5 Indigenous method of hard-variety *chhurpi* preparation in Sikkim and Bhutan.

4.1.5 Chhurpi (soft variety)

Soft-variety *chhurpi* or *kachcha chhurpi* (Pal et al. 1994) is a soft, cheese-like Himalayan fermented milk product. It has a rubbery texture with a slightly sour taste and aroma. It is consumed mostly in Sikkim, Darjeeling hills, and Bhutan (Figure 4.7).

4.1.5.1 Indigenous knowledge of preparation

Buttermilk is cooked for about 15 min until a soft and whitish mass is formed. The mass is sieved out and put inside a muslin cloth, which

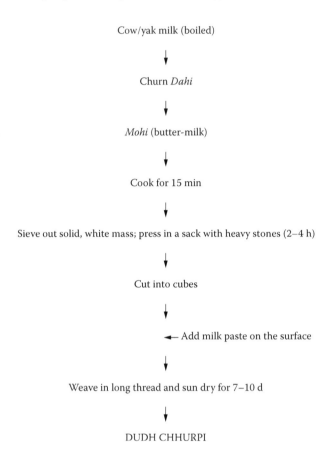

Cow/yak milk (boiled)

↓

Churn *Dahi*

↓

Mohi (butter-milk)

↓

Cook for 15 min

↓

Sieve out solid, white mass; press in a sack with heavy stones (2–4 h)

↓

Cut into cubes

↓

◄— Add milk paste on the surface

↓

Weave in long thread and sun dry for 7–10 d

↓

DUDH CHHURPI

Figure 4.6 Indigenous method of *dudh chhurpi* preparation in Sikkim.

is hung by a string to drain out the remaining whey. The product is the soft variety of *chhurpi* (Figure 4.8). Freshly prepared *chhurpi* can be kept at room temperature for 3–4 days during summer or 6–8 days in winter; otherwise, it becomes more sour and unacceptable to consumers.

4.1.5.2   Culinary practices and economy

Soft-variety *chhurpi* is prepared into a curry by cooking in edible oil or *gheu* along with onions, tomato, and chillies and is eaten with boiled rice. The curry is also prepared with wild edible ferns (*Diplazium esculentum*). It is also used to prepare *achar* or pickle by mixing it with chopped cucumber, radish, chillies, etc. (Tamang and Tamang 1998) (see Figure 4.9). Soup is also prepared from *chhurpi*.

Chhurpi is sold in local markets by women. It is packed in the leaves of the fig (*Ficus* sp.) plant and then tied loosely by straw. One kg of *chhurpi* costs about Rs. 60 or more.

Figure 4.7 Soft-*chhurpi* of Darjeeling hills.

4.1.5.3   Microorganisms

Lactic acid bacteria (LAB): *Lactobacillus plantarum, Lb. curvatus, Lb. fermentum, Lb. paracasei* subsp. *pseudoplantarum, Lb. alimentarius, Lb. kefir, Lb. hilgardii, Enterococcus faecium,* and *Leuconostoc mesenteroides* (Tamang et al. 2000).

4.1.6   Chhu

Chhu or *sheden,* an ethnic Himalayan fermented milk product, is a strong-flavored traditional cottage cheese–like product prepared from cow or yak milk in Sikkim, Darjeeling hills, some parts of Arunachal Pradesh, Ladakh, the high mountains of Nepal, Bhutan, and China (Tibet).

4.1.6.1   Indigenous knowledge of preparation

Shyow (meaning "curd" in Tibetan language) is churned in a bamboo or wooden vessel, with addition of warm or cold water to produce *maa* and *kachhu*. The latter is cooked for 15 min until a soft and white mass is formed. This mass is sieved out and put inside a muslin cloth, which is hung by a string to drain out the remaining whey. The product is called *chhu* (Figure 4.10). *Chhu* is placed in a closed vessel and kept for several days or months to ferment the product further, after which it is consumed.

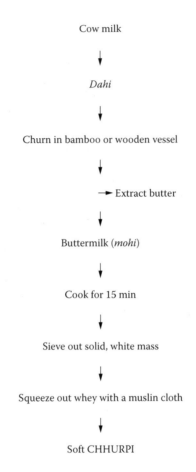

Cow milk

↓

Dahi

↓

Churn in bamboo or wooden vessel

↓

→ Extract butter

↓

Buttermilk (*mohi*)

↓

Cook for 15 min

↓

Sieve out solid, white mass

↓

Squeeze out whey with a muslin cloth

↓

Soft CHHURPI

Figure 4.8 Indigenous method of soft *chhurpi* preparation in Sikkim and Nepal.

4.1.6.2   Culinary practices

Fresh *chhu* is placed inside a vessel with a tight lid and left for fermentation at room temperature for a period ranging from a few days to several months. It is prepared into a curry by cooking in *maa* (butter) along with onions, tomato, and chillies, and mixed with salt. Soup prepared from strong-flavored *chhu* (kept for several months) is also consumed. It has a sour taste with a strong aroma and is used as an appetizer. *Ema dachi*, a hot curry prep-aration of *chhu*, is considered to be a most delicious food in Bhutan.

4.1.6.3   Microorganisms

Lactic acid bacteria (LAB): *Lactobacillus farciminis, Lb. brevis, Lb. alimentarius, Lb. salivarius, Lactococcus lactis* subsp. *cremoris, Saccharomycopsis* spp., and *Candida* spp. (Dewan and Tamang 2006).

Figure 4.9 *Chhurpi* curry.

4.1.7   Somar

Somar is an ethnic Himalayan fermented milk product consumed by the Sherpa tribes of Nepal and Sikkim, living in the high mountains. It is prepared from cow or yak milk. *Somar* is a soft paste with a bitter taste and has a strong flavor (Figure 4.11).

4.1.7.1   Indigenous knowledge of preparation

Buttermilk (*tara* in Sherpa dialect) is cooked until a soft, whitish mass is formed, and the mass is sieved out with a cloth or plastic sieve. The product is called *sherkam* (same as fresh soft *chhurpi*) and is placed in a closed vessel and kept for 10–15 days to ferment the product further. The final fermented product is called *somar*. In another traditional way, *somar* is cooked with milk, *mar* (butter), and turmeric to produce a soft brown paste *somar* (Figure 4.12). This type of *somar* can be stored for 4–7 months for future consumption

4.1.7.2   Culinary practices

A little oil is heated in a vessel; *somar* is added to it along with garlic, onion or *timbur* (*Zanthoxylum nitidum*), and salt. Water is added to produce a thin

Cow/yak milk (boiled or unboiled)

↓

Shyow (curd)

↓

Churn in bamboo or wooden vessel

↓

Kachhu (buttermilk)

↓

Cook for 15 min

↓

Sieve out solid, white mass

↓

Squeeze out whey with a muslin cloth

↓

Fresh *chhu*

↓

Ferment (7–30 d)

↓

CHHU

Figure 4.10 Indigenous method of *chhu* preparation in Sikkim.

soup. *Somar* soup, which is very tasty, is consumed with rice or *dheroh*. *Somar* is believed to have strong therapeutic value. It is used to cure stomach upset and acute diarrhea.

Somar is not sold in the local markets; it is usually prepared for home consumption.

4.1.7.3   Microorganisms
Lactic acid bacteria (LAB): *Lactobacillus paracasei* subsp. *pseudoplantarum* and *Lactococcus lactis* subsp. *cremoris* (Tamang et al. 2004).

Figure 4.11 *Somar* of Darjeeling hills.

4.1.8 Philu

Philu is a fermented, creamlike milk product obtained from cow or yak milk. It is commonly eaten by the Tibetans, Bhutia, Drukpa, and Sherpa.

4.1.8.1 Indigenous knowledge of preparation

Fresh milk—collected in cylindrical bamboo vessels (called *dzydung* by the Bhutia) or in wooden vessels (called *yadung*)—is slowly swirled around the walls of the vessels by rotating them for 10 min. Sometimes a thick mesh of dried creeper is kept inside the vessel to increase the surface area for the *philu* to stick (Figure 4.13). A creamy mass sticks to the walls of the vessels and around the creeper. The milk is then poured off and utilized elsewhere. The vessel is then kept upside down to drain out the remaining liquid. This process is repeated daily for about 6–7 days until a thick, white, creamy layer is formed on the vessel walls and the creeper surface (Figure 4.14). The soft-mass *philu* is scraped off and stored in a dry place for later consumption. *Philu* produced from cow milk is white colored, with a butterlike texture and slightly bland taste, while *philu* produced from yak milk has a creamy white color with an inconsistent semisolid texture.

4.1.8.2 Culinary practices and economy

Philu is cooked with butter and salt is added to make gravy, and it is eaten as a side dish along with boiled rice. Sometimes, it is mixed with meat and

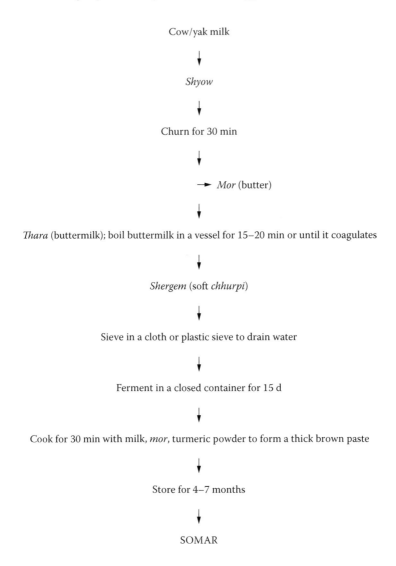

Figure 4.12 Indigenous method of *somar* preparation in Sikkim and Nepal.

vegetables. *Philu* is an expensive ethnic milk product sold in local markets in the Himalayas, costing Rs. 200 per kilogram.

4.1.8.3   Microorganisms
Lactic acid bacteria (LAB): *Lactobacillus paracasei* subsp. *paracasei, Lb. bifermentans,* and *Enterococcus faecium* (Dewan and Tamang 2007).

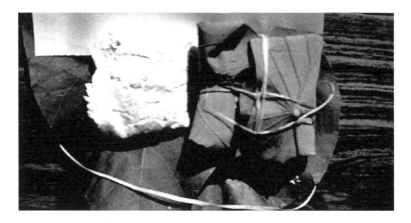

Figure 4.13 *Philu* of Sikkim.

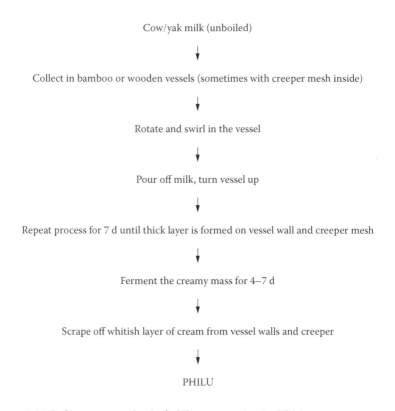

Cow/yak milk (unboiled)

↓

Collect in bamboo or wooden vessels (sometimes with creeper mesh inside)

↓

Rotate and swirl in the vessel

↓

Pour off milk, turn vessel up

↓

Repeat process for 7 d until thick layer is formed on vessel wall and creeper mesh

↓

Ferment the creamy mass for 4–7 d

↓

Scrape off whitish layer of cream from vessel walls and creeper

↓

PHILU

Figure 4.14 Indigenous method of *philu* preparation in Sikkim.

4.1.9   Kalari

Kalari is an ethnic fermented cheeselike product consumed by the Gujjars of Jammu & Kashmir (Katiyar et al. 1989). A similar product, called *paneer,* is a common dairy product in the cuisine of the western Himalayan people.

4.1.9.1   Indigenous knowledge of preparation
It is prepared by coagulating the cow milk at 50°C–60°C by cold whey, from which water is removed by straining through cheesecloth. It is then pressed into a disc shape. It is allowed to ripen, whereby it develops a typical pleasing aroma. The shelf life of *kalari* is 6–7 days at room temperature.

4.1.9.2   Culinary practices
Kalari is consumed in the Jammu region after being fried in hot edible oil. *Paneer* is fried and made into curry and side dishes.

4.1.9.3   Microorganisms
Lactic acid bacteria (LAB): unknown.

4.2   Microbiology

Lactic acid bacteria are the dominant microorganisms, with a load of 10^8 cfu/g in the Himalayan fermented milk products (Dewan and Tamang 2007). Lactic rods are represented by 86%, whereas cocci are represented by only 14% in the Himalayan fermented milk products. Rods dominate the lactic acid microflora in the Himalayan fermented milk products (Dewan 2002). This finding appears to be novel and has not yet been reported in the literature. Homofermentative lactics consist of *Lactobacillus plantarum, Lb. alimentarius, Lb. casei* subsp. *pseudoplantarum, Lb. casei* subsp. *casei, Lb. farciminis,* and *Lb. salivarius;* heterofermentative lactics include *Lb. bifermentans, Lb. hilgardii, Lb. kefir, Lb. brevis,* and *Lb. confuses;* cocci strains of LAB consist of *Lactococcus (Lc.) lactis* subsp. *lactis, Lc. lactis* subsp. *cremoris,* and *Enterococcus faecium.* The importance of LAB in the fermentation process of the Himalayan fermented milks is indicated by their numbers exceeding 10^7 cfu/g, based on the dilution factors. Microorganisms, mostly LAB, present in raw cow milk may contribute to the spontaneous fermentation of the milk (Gaya et al. 1999; Gadaga et al. 2001). The identity of the LAB isolated from the Himalayan fermented milks corresponds with that of LAB typically reported from other dairy products (Hammes and Vogel 1995). Spore-forming rods are also isolated from some of the fermented milk products and are identified as *Bacillus subtilis.*

It is interesting to observe that rods are only detected in *philu,* made from yak milk, suggesting that the milk type is the main driver for the flora present in *philu* and that yak milk products may have a distinctive flora.

Species of LAB recovered in the Himalayan milk products correspond to those of LAB typically reported for dairy products of other geographical regions (Gran et al. 2003; Mathara et al. 2004; Chammas et al. 2006).

The presence of yeasts (10^3 to 10^6 cfu/g) in the Himalayan fermented milk products indicate some role for this group, especially during the early stages of traditional and partly spontaneous fermentation processes and also flavor development (Dewan 2002). This observation may be justified by similar work on cheddar cheese, where yeasts have roles in aging and flavor development (Welthagen and Viljoen 1998). Lactose-fermenting *Candida* is isolated from *mohi* and *philu*. There are reports of lactose and nonlactose fermenting yeasts isolated from milk and fermented milk products (Ashenafi 1989; Paula et al. 1998). Depending on the substrate and environmental conditions, yeasts may show up to 10-fold higher metabolic activities as bacteria and may particularly influence the quality and acceptability of the final product. Yeasts bring about desirable fermentation changes in some fermented milk products (Westall and Filtenborg 1998). Yeasts are important in ripening of certain cheeses (Olson 1996). Although many yeasts can cause spoilage in milk products (Westall and Filtenborg 1998), there is little information on their significant role in lactose fermentation in fermented milk.

4.2.1   Pathogenic contaminants

Food safety is an important consideration in consumption of fermented foods. *Bacillus cereus* has been detected in few samples of *chhu, somar, philu,* and *dudh chhurpi*. However, none of these fermented milk samples were found to contain more than 10^3 cfu/g of *Bacillus cereus* population (Dewan 2002). A small amount of *B. cereus* in foods is not considered significant (Roberts et al. 1996). *Staphylococcus aureus* has been detected in a few samples of soft-variety *chhurpi, chhu, somar, philu,* and *dudh chhurpi* at the level of 10^3 cfu/g. Enterobacteriaceae population has been detected in sample of the Himalayan fermented foods below the level of 10^3 cfu/g (Dewan 2002). Occurrence of *B. cereus, S. aureus,* and Enterobacteriaceae in milk products is due to their presence in unsterilized milk. Raw milk (cow and yak) contains all pathogenic contaminants (Dewan and Tamang 2007). Lactic acid, produced by LAB, may reduce pH to a level where pathogenic bacteria (*Staphylococcus aureus, Bacillus cereus, Clostridium botulinum*) are inhibited or destroyed (Holzapfel et al. 1995). There has been no reported case of toxicity or illness due to consumption of the Himalayan fermented milk products.

4.2.2   Technological or functional properties

The acidifying and coagulating abilities of LAB strains are important technological properties (de Vuyst 2000; Herreros 2003). All LAB strains isolated from the Himalayan fermented milk products except *Lb. hilgardii* cause coagulation of milk at both 30°C and 37°C, with a significant drop in pH (Dewan and Tamang 2007). However, coagulation occurs faster at 37°C, ranging between 17–36 h, than at 30°C, with the coagulation period ranging between 20–38 h. Coagulation of milk by LAB strains shows their potential as starters or adjunct cultures in the production of the Himalayan fermented milk products.

The process of cheese maturation involves sequential breakdown of milk components, such as fat, protein, and lactose, by the starter bacteria (Davies and Law 1984). Therefore, a fundamental understanding of starter-culture enzymes is of prime importance in evaluating their suitability and in predicting their influence on the final cheese quality. The use of a commercial API-Zym test kit (bioMérieux, France) has been reported as a rapid and simple means of evaluating and localizing 19 different hydro-lases of microorganisms associated with dairy fermentations (Arora et al. 1990). Each of the LAB strains isolated from the Himalayan fermented milk products produces a wide spectrum of enzymes. They show rela-tively weak esterase and strong phosphatase, leucine-arylamidase, and β-galactosidase activities. However, they do not show detectable protei-nase activity. The absence of proteinases (trypsin and chymotrypsin) and presence of strong peptidase (leucine-arylamidase) and esterase-lipase (C4 and C8) activities produced by the LAB strains are traits of desirable quality for their use in production of typical flavor.

The API-zym technique may be used for selection of strains as poten-tial starter cultures on the basis of superior enzyme profiles, and also for accelerated maturation and flavor development of milk products (Tamang et al. 2000). The high activity of β-galactosidase exhibited by LAB species indicates that the ethnic Himalayan fermented milk products are suitable for consumption by lactose-intolerant persons. A decline in lactose content and an increase in lactase activity due to β-galactosidase activity of the starter culture make fermented milk more suitable for lactose-intolerant infants (Shah 2004). A similar observation was made in low-lactose *dahi*, which is found to be suitable for infants suffering from lactose-intolerance due to the presence of β-galactosidase (Chandrasekaran et al. 1975; Rao et al. 1985). Yogurt bacteria produce higher β-galactosidase than probiotic bacteria (Shah and Jelen 1990; Shah 1994, 2005). The high activity of phos-phatase by LAB strains showed their possible role in phytic acid degrada-tion in fermented milk products.

Antagonism refers to the inhibition of other (undesired or patho-genic) microorganisms, caused by competition for nutrients, and by the production of antimicrobial metabolites (Holzapfel et al. 1995). Strains of LAB compete with other microbes by screening antagonistic compounds and modifying the microenvironment by their metabolism (Lindgren and Dobrogosz 1990). Antagonistic properties of LAB strains were tested against *Listeria innocua* DSM 20649, *Listeria monocytogenes* DSM 20600, *Bacillus cereus* CCM 2010, *Staphylococcus aureus* S1, *Pseudomonas aeruginosa* BFE 162, *Enterobacter agglomerans* BFE 154, *Enterobacter cloacae* BFE 282, and *Klebsiella pneumoniae* subsp. *pneumoniae* BFE 147 (Dewan 2002). Most of the LAB strains isolated from the Himalayan fermented milk products show antimicrobial activities, mostly against these indicator bacteria (Dewan and Tamang 2007). LAB in *dahi* has shown antimicrobial activities (Balasubramanyam and Vardaraj 1994).

Biogenic amines have been reported in fermented milk products (Ten Brink et al. 1990; Halász et al. 1994). Lactic acid bacteria frequently produce histamine and tyramine in a variety of foods such as processed cheese, fish, and fermented vegetables and beverages (Stratton et al. 1991; Leisner et al. 1994). Strains of LAB isolated from the Himalayan fermented milks have been screened for their ability to produce biogenic amines. However, they did not produce biogenic amines in the applied method (Dewan and Tamang 2006, 2007). Most functional LAB do not produce significant levels of biogenic amines (Nout 1994). Nonproduction of biogenic amines by the LAB strains from the Himalayan milk products is a good indication of their acceptability for development of starter cultures. Production of biogenic amines by LAB strains is an undesirable property for selection of starter cultures (Buchenhüskes 1993). Before confirming the nonproduction of biogenic amine in the ethnic fermented milk products, qualitative and quantitative analyses of biogenic amine are necessary.

Some strains of LAB show high degrees of hydrophobicity (>75%), among which *Lb. paracasei* subsp. *paracasei* (isolated from *philu* made from cow milk) show the highest percentage of hydrophobicity of 98% (Dewan and Tamang 2007). Some strains of LAB have more than 75% hydrophobicity, indicating their probiotic character. For the first time, Tamang et al. (2000) reported the probiotic properties of microorganisms isolated from *chhurpi* samples.

4.2.3   In situ fermentation

For in situ fermentation studies, soft-variety *chhurpi* was selected because of its easy availability and preparation method (Dewan 2002). It was prepared in the laboratory following the traditional method, and changes in microbial load and accompanying physicochemical aspects were measured at 1-day intervals within a range of 0–6 days (Dewan 2002).

During soft-variety *chhurpi* fermentation, the population of LAB increased remarkably and reached the level of 10^8 cfu/g at the end of fermentation (Table 4.1). The load of lactic acid bacteria increased from 10^5 cfu/g in boiled milk to 10^8 cfu/g at the end of fermentation. Population of yeasts increased from 10^3 cfu/g to 10^7 cfu/g during fermentation. Subsequently, the load of *Bacillus* increased up to 10^3 cfu/g on the sixth

Table 4.1 Microbial Changes during Soft *Chhurpi*
Fermentation from Cow Milk

Fermentation Time (days)	Log cfu/g		
	LAB	Yeast	*Bacillus*
0_{un}[a]	7.8 (7.2–7.9)	7.0 (6.6–7.2)	2.0 (1.9–2.1)
0_b[b]	5.2 (4.8–5.6)	3.9 (3.7–4.1)	1.8 (1.7–2.0)
1	8.0 (7.8–8.2)	4.3 (4.0–4.6)	2.1 (1.8–2.3)
2	8.5 (8.3–8.7)	5.0 (4.8–5.3)	2.2 (2.0–2.4)
3	8.9 (8.6–9.0)	5.6 (5.3–5.8)	2.2 (2.1–2.4)
4	8.9 (8.6–9.1)	6.9 (6.7–7.1)	2.3 (2.0–2.5)
5	8.7 (8.5–8.8)	7.0 (6.8–7.2)	2.8 (2.5–3.0)
6	8.6 (8.5–8.7)	7.6 (7.5–7.7)	3.4 (3.2–3.7)

Note: Data represent the means of three batches of fermentation at
30°C. Ranges are given in parentheses.

[a] 0_{un} = Unboiled milk, 0 d.
[b] 0_b = Boiled milk, 0 d.
Source: Data from Dewan, 2002.

Table 4.2 Physicochemical Changes during Soft *Chhurpi*
Fermentation from Cow Milk

Fermentation time (days)	pH	Acidity (%)	Reducing Sugar (%)
0_{un}[a]	6.45 (6.42–6.46)	0.16 (0.14–0.18)	6.8 (6.7–6.9)
0_b[b]	6.38 (6.37–6.39)	0.18 (0.16–0.20)	6.6 (6.5–6.7)
1	4.28 (4.27–4.29)	0.59 (0.55–0.61)	5.4 (5.3–5.5)
2	4.22 (4.21–4.23)	0.61 (0.58–0.64)	1.7 (1.5–1.9)
3	4.20 (4.18–4.22)	0.62 (0.61–0.63)	0.7 (0.6–0.8)
4	4.17 (4.16–4.18)	0.66 (0.63–0.69)	0.6 (0.5–0.7)
5	4.13 (4.12–4.14)	0.75 (0.71–0.79)	0.5 (0.4–0.6)
6	4.08 (4.07–4.09)	0.82 (0.79–0.83)	0.4 (0.3–0.5)

Note: Data represent the means of three batches of fermentation at 30°C.
Ranges are given in parentheses.

[a] 0_{un} = unboiled milk, 0 d.
[b] 0_b = boiled milk, 0 d.
Source: Data from Dewan, 2002.

day. The mean pH value of boiled milk decreased from 6.4 to 4.1 on the
sixth day of fermentation (Table 4.2). Titratable acidity increased from 0 d
until the end of fermentation. Reducing sugar content decreased remarkably during fermentation. Due to lactic acid fermentation, pH decreased
(and acidity increased) during fermentation (Table 4.2). Reducing sugar
decreased gradually during fermentation due to amylolytic activity of
LAB, breaking glucose to lactic acid, thus decreasing pH.

4.3   Nutritive value

Nutritional composition of the Himalayan ethnic fermented milk (both cow and yak) products is presented in Tables 4.3 and 4.4 (Dewan 2002). The pH of all these products is acidic in nature, with higher acidity content due to lactic acid fermentation. A high content of protein and fat

Table 4.3 Nutritional Composition of the Himalayan Fermented Milk Products Prepared from Cow Milk

			Parameter					
			%		% (on dry matter basis)			Food Value (kcal/
Product	pH	Acidity	Moisture	Ash	Fat	Protein	Carbohydrate	100 g DM)
Cow milk	6.5	0.15	87.0	5.5	30.8	28.0	35.8	531.9
Dahi	4.2	0.73	84.8	4.7	24.5	22.5	48.2	503.6
Mohi	3.9	0.73	92.6	2.7	12.4	44.7	40.2	451.2
Soft *chhurpi*	4.3	0.61	73.8	6.6	11.8	65.3	16.3	432.4
Dudh chhurpi	6.0	0.29	16.8	5.2	6.1	57.2	31.6	409.4
Chhu	6.3	0.15	75.5	1.9	5.8	58.4	33.9	421.1
Somar	6.0	0.04	36.5	2.7	15.4	35.0	46.9	465.7
Philu	4.3	0.61	38.2	3.6	52.0	32.0	12.5	645.6

Note: Data represent the means of five samples.
Source: Data from Dewan, 2002.

Table 4.4 Nutritional Composition of the Himalayan Fermented Milk Products Prepared from Yak Milk

			Parameter					
			%		% (on dry matter basis)			Food Value (kcal/
Product	pH	Acidity	Moisture	Ash	Fat	Protein	Carbohydrate	100 g DM)
Yak milk	6.7	0.16	84.2	5.8	61.7	26.2	6.3	685.1
Dahi	4.0	0.82	86.3	7.5	69.4	28.2	5.1	758.0
Mohi	3.7	0.89	90.7	7.7	14.1	59.2	19.0	439.4
Maa	5.9	0.28	12.6	1.1	96.1	1.6	1.2	876.3
Chhu	4.8	0.44	70.1	6.1	11.2	62.5	20.2	431.6
Philu	4.8	0.79	47.9	1.4	48.0	38.7	12.0	634.5

Note: Data represent the means of five samples. Ranges are given in parentheses.
Source: Data from Dewan, 2002.

Table 4.5 Mineral Contents of the Himalayan Fermented Milk Products

Product	mg/100 g				
	Calcium	Iron	Magnesium	Manganese	Zinc
Cow soft *chhurpi*	44.1	1.2	16.7	0.6	25.1
Cow *chhu*	111.0	4.5	64.3	3.1	87.6
Cow *somar*	31.2	0.4	13.7	0.5	17.2
Cow *philu*	34.9	0.8	16.9	0.9	27.1
Cow *dudh chhurpi*	19.8	0.5	6.3	0.4	10.0
Yak *maa*	81.2	1.0	32.4	1.8	43.9

Note: Data represent the means of two samples.

is observed in all milk products. Ethnic fermented milk products are high-calorie-content foods of which *maa* (butter) made from yak has a high calorie value of 876.3 kcal/100 g. Among the minerals of the milk products, calcium content is higher than other minerals estimated (Table 4.5). Yak-sourced hard varieties of *chhurpi* are nutritious and give extra energy as a source of protein and fat (Katiyar et al. 1989) to the people living in snowbound areas. Probiotic properties are observed in many ethnic Himalayan fermented milk products (Tamang et al. 2007a). *Kalari* contains 37.8% moisture, 2.5% ash, 18% protein, 39% fat, 3.4% carbohydrate, potassium 325 mg/100 g, and calcium 115 mg/100 g (Katiyar et al. 1989).

4.4   Conclusion

Ethnic fermented milk products are traditionally produced at home in the Himalayas. Besides, *dahi, gheu, mohi, lassi, kalari,* and *paneer,* the rest of the fermented ethnic fermented milk products of the Himalayas are mostly prepared and consumed by people of Tibetan origin in the high mountains. Traditionally, very few fermented milk products are consumed by the ethnic people of North East India, except for the Tibetan-origin ethnic people of Arunachal Pradesh. Some ethnic milk products, especially *chhurpi,* are highly nutritious and provide the required energy to people living in high mountain regions. The microbial diversity in the Himalayan fermented milk products—ranging from diverse species of LAB to yeasts—may contribute significant information on unknown microbial gene pools as genetic resources of the Himalayas.

chapter five

Fermented cereals

Consumption of rice and its fermented products is increasing in the developing countries due to its high caloric value, appealing quality characteristics, and high acceptability by consumers (Steinkraus 1994). Rice has long been a staple food in Asia, and this history has led to a wide variety of traditional cereal fermentations with molds and yeasts (Haard et al. 1999). Varieties of traditional nonalcoholic cereal-based fermented foods are mostly prepared and consumed in Africa as staple foods (Nout 2001; Blandino et al. 2003). Fermented cereal-based gruels are generally used as naturally fortified weaning foods for young children in Africa (Tou et al. 2007). Cereals are fermented either to produce alcoholic beverages and drink or to prepare varieties of bakery or staple nonalcoholic foods worldwide. Alcoholic food beverages from cereals such as rice, finger millets, barley, maize, etc., are prepared by using dry mixed cultures in Asian countries (discussed in Chapter 8). Ethnic nonalcoholic fermented cereal foods are mostly prepared and consumed as staple foods in the form of breads, loafs, confectionery, and gruels worldwide (Oyewole 1997) as well as complementary foods for infants and young children in Africa (Nout 1991; Efiuvwevwere and Akona 1995; Tou et al. 2006).

Several fermented cereal products across the world have been well investigated, including sourdough of America, Australia, and Europe (Brandt 2007); *masa* of South Africa (Efiuvwevwere and Ezeama 1996); *puto* of Southeast Asia (Kelly et al. 1995; Sanchez 1996); *mawé* or *ogi* of Benin (Onyekwere et al. 1989; Hounhouigan et al. 1993a); *kisra* of Sudan (Mohammed et al. 1991); *ben-saalga* of Burkino Faso (Tou et al. 2007); *kenkey* of Ghana (Nche et al. 1994; Nout et al. 1996); *togwa* of Tanzania (Mugula et al. 2003); *tarhana* of Turkey (Erbas et al. 2006); *idli* of India (Steinkraus et al. 1967; Reddy 1981; Soni and Sandhu 1991); *dosa* of India (Soni et al. 1985, 1986); and *rabadi* of India (Gupta et al. 1992a). The production of bread is an ancient biotechnological process that is based on the fermentation of a wheat-flour dough, but flours from other cereals such as rye are also used, and now it has evolved into a modern industrialized process (Jenson 1998; Hammes and Ganzle 2005; Decock and Cappelle 2005).

Various indigenous varieties of cereal crops are cultivated in the Himalayas, depending upon the elevations. These include rice (*Oryza sativa* L.), maize (*Zea mays* L.), finger millet (*Eleusine coracana* Gaertn.), wheat (*Triticum aestivum* L.), barley (*Hordeum vulgare* L.), buckwheat

(*Fagopyrum esculentum* Moench.), sorghum (*Sorghum vulgare* Pers.), pearl millet (*Pennisetum typhoides* L.), etc. Rice and maize are eaten mostly in the eastern Himalayas, whereas wheat is eaten as a staple food in the western Himalayas, and barley and finger millet are commonly eaten in the high mountain areas of the Himalayas. Cereals are mostly used as non-fermented staple foods and also for production of alcoholic beverages in the Himalayas. *Bhat*, cooked rice, is a staple food in the eastern Himalayas, whereas chapati or roti, made from wheat flour, is common in the western Himalayas. *Dheroh*, cooked maize, is a common staple in the diet of the poor rural people in Nepal and in some parts of Sikkim and Bhutan.

5.1   Important fermented cereals

Only a few cereals are traditionally fermented into the products that are special foods in the cuisines of the Himalayan people. These include *selroti* in Nepal, Darjeeling hills, and Sikkim; *nan* in the western Himalayas; *jalebi* throughout the entire Himalayas; and *siddu*, *bhatarua*, and *seera* in Himachal Pradesh and Uttarakhand.

5.1.1   Selroti

The Nepali communities of the Himalayan regions of India, Nepal, and South Bhutan prepare a cereal-based fermented food called *selroti*. It is a popular fermented rice product that is ring-shaped, spongy, dough-nut-like, and a deep-fried food. It is prepared for religious festivals and special occasions. *Selroti* is a Nepali word for a ring-shaped, rice-based bread (Yonzan and Tamang 2009). Different ethnic groups call it by various names: *selsoplay* by the Mukhia, *selgaeng* by the Tamang, *selpempak* by the Rai, etc.

5.1.1.1   Indigenous knowledge of preparation
The local variety of rice (*Oryza sativa* L.), called *attey*, is sorted, washed, and soaked in cold water overnight or 4–8 h at ambient temperature. Sometimes, milled rice is also used for *selroti* preparation. Water is then decanted from the rice by using a bamboo sieve, called *chalni*, and spread over a woven bamboo tray, called *naanglo*, and dried for 1 h. Soaked rice is pounded into a coarse powder using a wooden mortar and pestle, known as *okhali* and *mushli*, respectively. Larger particles of pounded rice flour are separated from the rest by winnowing in a bamboo tray. Then the rice flour is mixed with nearly 25% refined wheat (*Triticum aestivum* L.) flour, 25% sugar, 10% butter or fresh cream, and 2.5% spices/condiments—large cardamom (*Amomum subulatum* Roxb.), cloves (*Syzygium aromaticum* Merr.), coconut (*Cocos nucifera* L.), fennel (*Foeniculum vulgare* Mill.), nutmeg

(*Myristica fragrans* Houtt.), cinnamon (*Cinnamomum zeylanicum* Bl.), and small cardamom (*Elletaria cardamomum* Maton.)—are added to the rice flour and mixed thoroughly. Some people add a tablespoon of honey or an unripe banana or baking powder (sodium bicarbonate) to the mixture, depending on its quantity. Milk (boiled/unboiled) or water is added, and the mixture is kneaded into a soft dough and finally into a batter with easy flow. Batter is left to ferment naturally at ambient temperature (20°C–28°C) for 2–4 h during summer and at 10°C–18°C for 6–8 h during winter. The oil is heated in a cast-iron frying pan, locally called *tawa*. The fermented batter is squeezed by hand or *daaru*, deposited as a continuous ring onto hot edible oil, fried until golden brown, and then removed from the hot oil by a poker, locally called *jheer* or *suiro*, or by a spatula, locally called *jharna* (Figure 5.1).

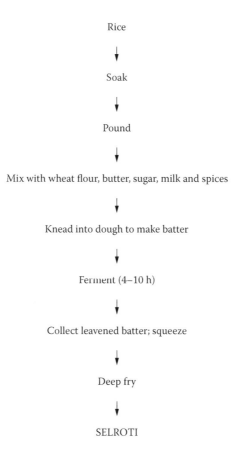

Rice

↓

Soak

↓

Pound

↓

Mix with wheat flour, butter, sugar, milk and spices

↓

Knead into dough to make batter

↓

Ferment (4–10 h)

↓

Collect leavened batter; squeeze

↓

Deep fry

↓

SELROTI

Figure 5.1 Indigenous method of *selroti* preparation in the eastern Himalayas.

5.1.1.2   Traditional equipment used during preparation of selroti
The following traditional equipment is used during preparation of *selroti*:

Okhali and *mushli*: A wooden mortar and pestle, respectively, used to pound soaked rice

Naanglo: A woven tray of bamboo strips used to dry soaked rice

Chalni: A sieve made of metal wire or bamboo strips; the bamboo sieve has bigger holes to drain water from soaked rice; the sieve made of metal wire is finely woven and is used to sieve pounded rice

Suiro: A pointed bamboo stick used to turn *selroti* upside down and then to lift and remove the fried *selroti* from the oil

Jheer: A poker made of metal wire to remove the deep-fried *selroti* from the hot oil

Daaru: A metallic serving spoon used to pour batter onto the hot edible oil (Figure 5.2)

Tawa: A cast-iron frying pan used to fry *selroti*

Jharna: A metal spatula that has a wide flat blade with holes; used to drain oil from the fried *selroti*

Thumsey: A bamboo basket use to store freshly fried *selroti* (Figure 5.3)

Figure 5.2 *Selroti* batter is deep-frying in hot edible oil.

Figure 5.3 Fried *selroti*.

5.1.1.3   Culinary practices

Selroti is served as a confectionary bread with *aalu dam* (boiled potato curry), *simi ko acchar* (pickle prepared from string beans), and meat in the Nepali meal. It can be served hot or cold. *Selroti* can be stored at room temperature for two weeks

5.1.1.4   Socioeconomy

The preparation of *selroti* is an art of technology and is a family secret passed from mother to daughter. Women prepare it, and men help them in pounding the soaked rice. Survey results indicated that *selroti* is mostly prepared at home. It is also sold in canteens, local food stalls, and restaurants. Some people are economically dependent upon this product. In local food stalls in Sikkim and Darjeeling hills, it is sold at the average rate of Rs. 10 per plate, each plate containing four pieces of fried *selroti*.

5.1.1.5   Microorganisms

Lactic acid bacteria (LAB): *Leuconostoc mesenteroides*, *Enterococcus faecium*, *Pediococcus pentosaceus*, and *Lactobacillus curvatus*; yeasts: *Saccharomyces cerevisiae*, *S. kluyveri*, *Debaryomyces hansenii*, *Pichia burtonii*, and *Zygosaccharomyces rouxii* (Yonzan 2007).

5.1.2   Jalebi

Jalebi is a crispy, sweet, deep-fried doughnut made from wheat flour and eaten as a snack food in the western Himalayas (Batra and Millner 1974; Chitale 2000).

5.1.2.1   Indigenous knowledge of preparation

Jalebi is prepared by mixing wheat flour (*maida*) with *dahi* (curd), adding water to it, and leaving overnight at room temperature. The thick leavened batter is squeezed through an embroidered hole (about 4 mm in diameter) in thick and durable cotton cloth, deposited as continuous spirals into hot edible oil, and fried on both sides until golden and crisp. After about a minute, these are removed from fat with a sieved spatula and then sub-merged for several seconds in a hot sugar saffron-scented syrup, which saturates their hollow insides. Skill is required to master the uniform shapes of *jalebi*. Often rose water or *kewda* (*Pandanus tectorius*) water and orange food color are added to the syrup (Ramakrishnan 1979; Batra 1981) (Figures 5.4 and 5.5).

5.1.2.2   Culinary practices

Jalebi is a dessert with impressive shapes, bright yellow-to-orange color, and intriguing textures. It is eaten as a snack when hot. It is a very popular

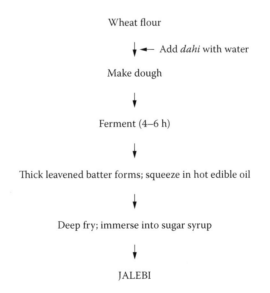

Figure 5.4 Indigenous method of *jalebi* preparation in Uttarakhand.

Figure 5.5 Deep-fried *jalebi.*

sweet dish in the western Himalayas, including Himachal Pradesh, Jammu & Kashmir, Uttarakhand, and Nepal. *Jalebi* is sold in all local sweet shops, and most vendors sell them while hot.

5.1.2.3   Microorganisms

Lactic acid bacteria (LAB): *Lactobacillus fermentum, Lb. buchneri, Lb. bulgaricus, Streptococcus lactis, S. thermophilus,* and *Enterococcus faecalis*; yeasts: *Saccharomyces bayanus, S. cerevisiae,* and *Hansenula anomala* (Batra and Millner 1976; Soni and Sandhu 1990b).

5.1.3   Nan

Nan is an indigenous leavened bread made from wheat flour and is commonly consumed by the ethnic people of the western Himalayas.

5.1.3.1   Indigenous knowledge of preparation

During preparation of *nan*, wheat flour (called *maida*) is thoroughly mixed with *ghee* (butter), baking powder, *dahi* (curd), milk, salt, sugar, and water

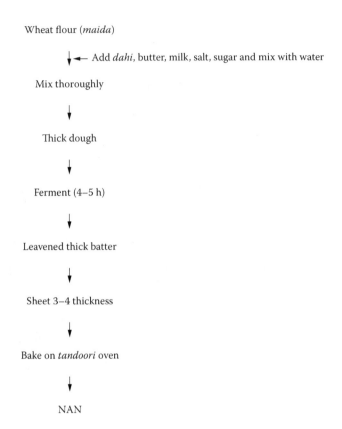

Wheat flour (*maida*)

↓ ← Add *dahi*, butter, milk, salt, sugar and mix with water

Mix thoroughly

↓

Thick dough

↓

Ferment (4–5 h)

↓

Leavened thick batter

↓

Sheet 3–4 thickness

↓

Bake on *tandoori* oven

↓

NAN

Figure 5.6 Indigenous method of *nan* preparation in Kashmir.

to make a thick dough. The dough is fermented for 3–5 h at room temperature. After this, the batter or dough is sheeted between the palms of the hand to about 2–3 mm thickness, wetted a little, and pasted on the inner wall of a tandoori oven. The baked product, called *nan*, has a typical soft texture and flavor (Figure 5.6).

5.1.3.2   Culinary practices
Nan is baked in a specially designed oven known as a tandoori within a temperature range of 300°C–350°C. It is also baked on live coal or flame for a short time of 2–3 min. *Nan* is eaten as a staple food with vegetable or *dal* (legume soup) and with meats in Kashmir and Himachal Pradesh of the western Himalayas.

5.1.3.3   Microorganisms
Yeast: *Saccharomyces kluyveri* (Batra 1986).

5.1.4 *Siddu*

Siddu, also called *khobli*, is an ethnic fermented wheat product commonly consumed in the Kullu and Shimla districts of Himachal Pradesh (Thakur et al. 2004).

5.1.4.1 *Indigenous knowledge of preparation*

During preparation of *siddu*, wheat flour is mixed with water, and *malera* (previously leftover dough of *siddu*) is added as a starter and the dough is kneaded. It is left for fermentation for 4–5 h at room temperature. Fermented dough gives an oval or disc shape. It is stuffed with spices, paste of opium seeds/walnut/black gram, and is steamed before consuming (Figure 5.7).

5.1.4.2 *Culinary practices*

Siddu is served hot with *ghee* (butter) or *chutney* (pickle) in rural areas of Himachal Pradesh as a special/occasional dish. It is also served as a traditional local dish in many restaurants in Kullu and Manali of Himachal Pradesh.

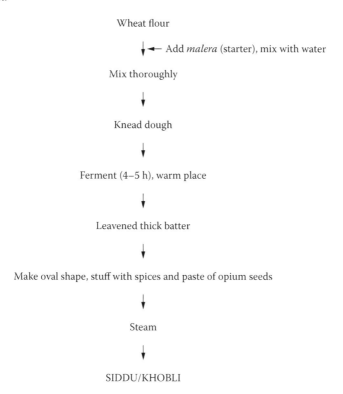

Wheat flour

↓ ← Add *malera* (starter), mix with water

Mix thoroughly

↓

Knead dough

↓

Ferment (4–5 h), warm place

↓

Leavened thick batter

↓

Make oval shape, stuff with spices and paste of opium seeds

↓

Steam

↓

SIDDU/KHOBLI

Figure 5.7 Indigenous method of siddu preparation in Himachal Pradesh.

5.1.4.3   Microorganisms

Lactic acid bacteria (LAB: unknown) and yeasts (unknown).

5.1.5   Chilra

Chilra, also called *lwar*, is fermented buckwheat or barley of Himachal Pradesh. It is similar to South Indian *dosa* (Savitri and Bhalla 2007).

5.1.5.1   Indigenous knowledge of preparation

During the indigenous method of *chilra* preparation in Himachal Pradesh, barley or wheat flour and buckwheat flour are mixed in the ratio of 1:3 with *treh* (previously leftover wheat flour slurry used as a starter) and water to make a thick slurry. It is fermented at room temperature for 10–12 h. Fermented slurry is speared on a hot plate and baked on both sides to serve *chilra* (Figure 5.8).

5.1.5.2   Culinary practices

Chilra is a staple food as well as a popular snack served with coriander pickle, baked potato, and mutton soup in Lahaul in Himachal Pradesh. It is also prepared for marriage ceremonies and festivals.

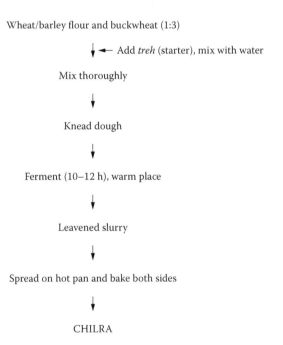

Wheat/barley flour and buckwheat (1:3)

← Add *treh* (starter), mix with water

Mix thoroughly

Knead dough

Ferment (10–12 h), warm place

Leavened slurry

Spread on hot pan and bake both sides

CHILRA

Figure 5.8 Indigenous method of *chilra* preparation in Himachal Pradesh.

5.1.5.3 Microorganisms

Lactic acid bacteria (LAB): *Lactobacillus, Leuconostoc,* and *Lactococcus* spp.; yeasts: *Saccharomyces, Debaryomyces,* and *Schizosaccharomyces* spp. (Kanwar et al. 2007).

5.1.6 Marchu

Marchu is a fermented bread or roti prepared from wheat flour in Himachal Pradesh (Thakur et al. 2004).

5.1.6.1 Indigenous knowledge of preparation

During *marchu* preparation, wheat flour is mixed with water, spices, and a little salt, and a dough is made. Starter culture, called *malera* (previously leftover dough), is added to the dough and is fermented for 10–12 h overnight. Fermented dough is deep-fried in mustard oil to make *marchu* roti (Figure 5.9).

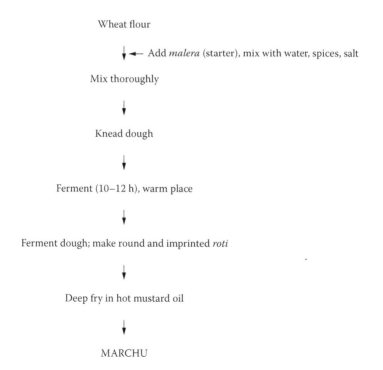

Figure 5.9 Indigenous method of *marchu* preparation in Himachal Pradesh.

5.1.6.2   Culinary practices

Marchu is eaten as a snack or breakfast with tea in Himachal Pradesh. It is prepared in Lahaul for local festivals (*phagli, halda*) and for religious and marriage ceremonies.

5.1.6.3   Microorganisms

Lactic acid bacteria (LAB: unknown) and yeasts (unknown).

5.1.7   Bhaturu

Bhaturu, also called *sumkeshi* roti, is an ethnic bread or roti prepared from fermented wheat flour or barley in Himachal Pradesh (Thakur et al. 2004).

5.1.7.1   Indigenous knowledge of preparation

During *bhaturu* preparation, wheat/barley flour is mixed with water and *malera*, and a dough is made. It is kept for fermentation for 3–5 h in a warm place. Fermented dough is made into a round-shaped roti and baked on an earthen oven for 20–30 min to make *bhaturu* (Figure 5.10).

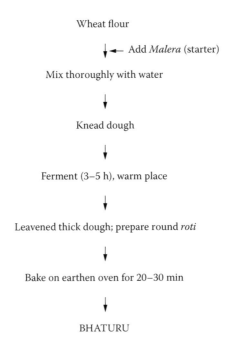

Wheat flour

↓◄— Add *Malera* (starter)

Mix thoroughly with water

↓

Knead dough

↓

Ferment (3–5 h), warm place

↓

Leavened thick dough; prepare round *roti*

↓

Bake on earthen oven for 20–30 min

↓

BHATURU

Figure 5.10 Indigenous method of *bhaturu* preparation in Himachal Pradesh.

5.1.7.2 Culinary practices

Bhaturu is a popular bread or roti with vegetables or curries. It serves as a diet staple for the rural people of Himachal Pradesh and even in Uttarakhand.

5.1.7.3 Microorganisms

Lactic acid bacteria (LAB): *Lactobacillus, Leuconostoc,* and *Lactococcus* spp.; yeasts: *Saccharomyces, Debaryomyces,* and *Schizosaccharomyces* spp. (Kanwar et al. 2007).

5.2 Microbiology

5.2.1 Selroti

The microbial population of *selroti* batters comprises LAB population in viable numbers above 10^8 cfu/g, followed by yeasts at around 10^5 cfu/g, respectively (Yonzan 2007). Taxonomically diverse species of LAB have been identified from *selroti* batters, including *Leuconostoc mesenteroides, Enterococcus faecium, Pediococcus pentosaceus,* and *Lactobacillus curvatus* (Yonzan 2007). *Leuconostoc mesenteroides* has been reported in several fermented cereal foods such as *idli* of India (Mukherjee et al. 1965), *enjera* of Ethiopia(Oyewole 1997), *puto* of the Philippines (Kelly et al. 1995), and *mawè* of Togo and Benin (Hounhouigan et al. 1993b). Enteroccoci play a beneficial role in the production of many fermented foods (Bouton et al. 1998; Cintas et al. 2000). *Enterococcus faecium* appears to pose a low risk for use in foods, because these strains generally harbor fewer recognized virulence determinants than *E. faecalis* (Franz et al. 2003). *Pediococcus pentosaceous* along with several *Lactobacillus* spp. are reported as predominant LAB strains in Tanzanian *togwa* (Mugula et al. 2003) and in *mawè* of Togo and Benin (Hounhouigan et al. 1993b). *Lactobacillus curvatus* has been reported in *mawè*, a fermented maize food, at a lower percentage of prevalence than other LAB (Hounhouigan et al. 1993c).

Along with LAB populations, yeasts have also been isolated and identified from *selroti* batters, including *Saccharomyces cerevisiae, Saccharomyces kluyveri, Debaryomyces hansenii, Pichia burtonii,* and *Zygosaccharomyces rouxii* (Yonzan 2007). *Saccharomyces cerevisiae* is the principal yeast of most bread fermentations (Jenson 1998) and sourdough production (Hammes and Ganzle 1998). Yeasts mostly produce gas, causing expansion and leavening of the dough, ultimately affecting the texture, density, and volume of the bread (Hammes et al. 2005). *Debaryomyces hansenii* has been isolated from *idli* along with several other yeasts (Soni and Sandhu 1991). *Pichia burtonii* has been reported in some Asian rice-based alcoholic starters such as *loog-pang* of Thailand (Limtong et al. 2002) and *marcha* of Sikkim

(Tsuyoshi et al. 2005). However, *Pichia burtonii* produces visible, white, or chalky discoloration in sourdough (Legan and Voysey 1991). The recovery of *P. burtonii* in *selroti* batter is possibly through the use of rice grains, as these contain the yeast *P. burtonii* (Kreger-van Rij 1984) and bacterial leuconostocs and pediococci (Wood and Holzapfel 1995). Acid-tolerant *Zygosaccharomyces rouxii* has not been reported in fermented cereal foods, although it has been reported in many fermented soybean foods of Asia, where it is valued for its contribution to aroma (Aidoo et al. 2006). The origin of *Z. rouxii* is usually from sugar, honey, and confectionery (Kreger-van Rij 1984). The recovery of *Z. rouxii* in *selroti* batters is probably because of its entry through the sugars and honey that are added during *selroti* batter preparation to make it sweet.

The prevalent LAB in samples of *selroti* batters collected from different sources in Sikkim and Darjeeling hills are *Leuc. mesenteroides* at 42.9%, followed by *P. pentosaceus* at 23.8%, *E. faecium* at 20.4%, and *Lb. curvatus* at 13.0%. Similarly, the most dominant yeasts recovered in all samples of *selroti* batters are *S. cerevisiae* at 35.6%, followed by *D. hansenii* at 17.6%, *P. burtonii* at 17.1%, *Z. rouxii* at 16.3%, and *S. kluyveri* at 13.4% (Yonzan 2007). The most dominant functional LAB and yeasts in *selroti* batters are *Leuc. mesenteroides* and *S. cerevisiae*, respectively (Yonzan 2007). The predominance of *Leuc. mesenteroides* and *S. cerevisiae* is also common in other fermented cereal foods (Steinkraus 1996; Brandt 2007).

5.2.1.1 Food safety

Food-borne pathogens *Bacillus cereus*, *Listeria* spp., *Salmonella* spp., and *Shigella* spp. have not been detected in batters of *selroti* due to the slightly acidic nature of the products prior to frying (Yonzan 2007). Fermentation of cereals reduces contamination of weaning foods in Ghana (Mensah et al. 1990). The chance of contamination during handling of raw materials for preparation is less, as the pH of the materials is low enough to inhibit the growth of pathogens. The high population (>10^8 cfu/g) of LAB in *selroti* batters could restrict the growth of other organisms. It is also a fact that lactic acid produced by LAB may reduce pH to a level where pathogenic bacteria may be inhibited (Holzapfel et al. 1995; Tsai and Ingham 1997; Adams and Nout 2001).

5.2.1.2 Effect of seasonal variation on microbial load

The effect of seasonal variation on the microbial load of *selroti* batters has recently been studied. Samples were prepared during summer and winter seasons in different restaurants, food stalls, canteens, and street vendors located in Sikkim (Yonzan 2007). During summer, the microbial loads of LAB and yeast were found to be 10^8 cfu/g and 10^4 cfu/g, respectively. In winter, the population LAB was counted at about 10^7 cfu/g, and the population of yeast was 10^6 cfu/g. The microbial load of LAB increases during summer due to the rise in room temperature, which may accelerate the fermentation rate, whereas winter favors the growth of yeasts. It was observed that the seasons did affect the development and prevalence of microorganisms in the fermented batters of *selroti*. A similar observation on the effect of seasonal variation on bacterial load was made during *idli* fermentation (Soni et al. 1986).

5.2.1.3 Technological properties

LAB and yeast strains isolated from *selroti* batters were screened for their acidifying and coagulating capacity, and it was found that most of the LAB strains acidified, with pH decreasing to as low as 4.3. About 63.6% of LAB strains cause coagulation of skim milk at 30°C. Coagulation of milk by LAB strains shows their potential as starters or adjunct cultures in the production of fermented products. Among the yeast strains, only *S. cerevisiae* and *D. hansenii* showed any acidification characteristics, although the decrease in pH was limited to 5.6.

The study of enzymatic properties in fermented products is important (Tamang et al. 2000; Kostinek et al. 2005). The absence of proteinases (trypsin) and the presence of strong peptidase (leucine-, valine-, and cystine-arylamidase) activities produced by the predominant LAB strains isolated from *selroti* batters are possible traits of desirable quality for their use in production of typical flavor and aroma. Yeasts in *selroti* show a high activity of phosphatase, advocating their phytic acid–degrading abilities in the product (Yonzan 2007). Antinutritive factors such as phytic acids and oligosaccharides are of particular significance in unbalanced cereal-based diets (Fredrikson et al. 2002). Due to these nutritional consequences, the degradation of antinutritive factors in food products by fermentation is desirable, as reported for a number of foods of plant origin (Chavan and Kadam 1989; Mbugua et al. 1992; Svanberg et al. 1993). *Leuconostoc mesenteroides* isolated from *selroti* batters has high α-galactosidase and β-galactosidase activities. The presence of high activity of α-galactosidase probably indicates an ability to hydrolyze oligosaccharides of the raffinose family (Holzapfel 2002).

All strains of LAB from batters of *selroti* show antimicrobial activities against a number of pathogenic Gram-positive and Gram-negative bacteria (Yonzan 2007). However, the cell-free supernatant fluid extracts of LAB strains isolated from *selroti* batters do not produce bacteriocin under the applied conditions. Species of LAB strains isolated from several fermented cereal foods have such antimicrobial activities, including production of bacteriocins (Olukoya et al. 1993; Omar et al. 2006).

5.2.1.4 In situ fermentation dynamics

During in situ fermentation of *selroti* batter, populations of LAB and yeasts change spontaneously, with a decrease in pH and an increase in the acidity of the fermenting substrates due to growth of LAB. Bacterial contaminants *Bacillus cereus* and Enterobacteriaceae are associated with initial fermentation and finally disappear during *selroti* batter fermentation. The LAB produced sufficient acid for inhibition of pathogenic microorganisms in foods (Adam and Nicolaides 1997). By averting the invasion of these potential contaminants, lactic acid fermentation imparts attributes of robust stability and safety in a product like *selroti*. Another safety aspect of *selroti* is deep frying prior to consumption. Various microorganisms occur at the early stages of *selroti* batter fermentation. This may be partly attributed to the microbial diversity often associated with rice grains and plant materials (Efiuvwevwere and Ezeama 1996). There is no remarkable increase in physical properties of fermenting cereal, such as batter temperature, volume, and weight, during in situ fermentation of *selroti* batter. However, increase in batter volume during fermentation has been reported in *idli* and *dosa* (Soni et al. 1986) and in *puto* (Tongananta and Orillo 1996). Numerous chemical and physical factors influence the rate and growth of various microorganisms, as well as their sequence of appearance during cereal fermentation (Cartel et al. 2007).

5.2.1.5   Selroti batter preparation by selected starter cultures
Starter cultures of LAB and yeasts previously isolated from native *selroti* batters were tested singly and in combination for their ability to ferment rice flour to produce *selroti*. The rationale behind the study was to use starter culture to supplement the natural microorganisms essential for leavening of *selroti* batter. Different starter cultures and their combinations were made:

Starter A: Cells of *Leuc. mesenteroides* BS1:B1
Starter B: Mixture of cells of all LAB strains (*E. faecium* BS1:B2; *Lb. curvatus* BP:B1; *Leuc. mesenteroides* BS1:B1; *P. pentosaceus* BG:B2)
Starter C: Cells of *S. cerevisiae* BA1:Y2
Starter D: Mixture of cells of all yeasts strains (*D. hansenii* BR1:Y4; *P. burtonii* BG1:Y1; *S. cerevisiae* BA1:Y2; *S. kluyveri* S3:Y3; *Zygosaccharomyces rouxii* S1:Y6)
Starter E: Mixture of starters B and D (LAB and yeasts)
Starter F: Mixture of cells of *Leuc. mesenteroides* BS1:B1 and *S. cerevisiae* BA1:Y2

Selroti was prepared in the laboratory following the traditional method (Yonzan 2007). About 350 g of batter was equally distributed in sterile 500-ml Duran bottles with screw caps. Each batter was inoculated with 1 ml of the starter culture(s) per 100 g of batter, either singly or in the combinations described in the previous list; mixed thoroughly with a

Table 5.1   Format for Sensory Evaluation of **Selroti**

Please use market Selroti as a control with scoring rate of 3 (moderate)

Sample Code:

Name:

Attribute	Score					Comment
Aroma	Bad				Excellent	
	1	2	3	4	5	
Taste	Bad				Excellent	
	1	2	3	4	5	
Texture	Hard				Soft	
	1	2	3	4	5	
Color	Bad				Excellent	
	1	2	3	4	5	
General acceptability	Bad				Excellent	
	1	2	3	4	5	

Date: Signature of the judge

Requirements for high-grade *Selroti*

Aroma:	Typical *Selroti* flavor
Taste:	Sweet
Texture:	Soft
Color:	Golden brown

Source: Yonzan, 2007.

sterile spatula; and incubated at 28°C. The pH and titratable acidity were determined at 0-, 4-, and 6-h intervals. *Selroti* prepared from diffcrent bat-ter samples incubated for 4 and 6 h were deep-fried in hot edible oil and served to consumers for sensory evaluation using the standard format (Yonzan 2007). Sensory evaluations (Table 5.1) were conducted to choose the best culture combinations. It was found that *selroti* batters produced using Starter-E, containing a mixture of pure culture strains of *Leuconostoc mesenteroides* BS1:B1 and *Saccharomyces cerevisiae* BA1:Y2, incubated at 28°C for 4 h, had the highest organoleptic score and thus the highest acceptabil-ity among the consumers. This result was also correlated by a decrease in pH (and concomitant increase in acidity) of the fermenting batters from 0 to 4 h.

Leuconostoc mesenteroides BS1:B1 was selected as a starter culture based on its heterofermentative property, superior technological properties

Table 5.2 Sensory Evaluation of *Selroti* Batter Prepared Using Selected Starter Cultures

Starter Culture	Fermentation Time (h)	Attribute				
		Aroma	Taste	Texture	Color	General Acceptability
A	4	3.0 ± 0.6^a	2.8 ± 0.7^b	2.7 ± 0.7^a	3.2 ± 0.7^a	2.9 ± 0.6^b
	6	2.7 ± 1.3^a	2.8 ± 1.1^b	3.7 ± 1.0^a	3.3 ± 0.7^a	2.8 ± 1.1^b
B	4	2.9 ± 0.6^a	2.8 ± 0.7^b	3.3 ± 0.8^a	2.2 ± 0.8^b	2.9 ± 0.6^b
	6	3.1 ± 0.6^a	2.9 ± 0.6^b	2.6 ± 1.0^a	3.1 ± 0.6^a	2.9 ± 0.6^b
C	4	3.0 ± 0.7^a	3.1 ± 0.7^a	3.8 ± 0.8^a	3.2 ± 0.3^a	3.2 ± 0.6^b
	6	2.7 ± 0.6^a	2.7 ± 0.7^b	2.6 ± 0.9^a	3.3 ± 0.6^a	2.9 ± 0.8^b
D	4	2.9 ± 0.6^a	2.9 ± 1.0^b	2.7 ± 0.8^a	2.2 ± 0.8^b	2.8 ± 0.8^b
	6	2.4 ± 0.6^a	2.9 ± 0.6^b	3.3 ± 0.8^a	3.2 ± 0.5^a	2.8 ± 0.8^b
E	4	2.5 ± 0.9^a	2.1 ± 1.0^b	1.9 ± 1.2^b	2.3 ± 0.9^b	2.1 ± 1.0^b
	6	2.7 ± 0.5^a	2.1 ± 1.2^b	2.0 ± 1.1^b	2.6 ± 0.7^a	2.2 ± 1.1^b
F	4	4.0 ± 1.0^a	4.6 ± 1.1^a	4.0 ± 1.1^a	3.9 ± 0.7^a	4.8 ± 1.0^a
	6	3.1 ± 1.0^a	3.4 ± 1.1^a	3.8 ± 1.1^a	3.7 ± 0.8^a	3.3 ± 1.3^b
G	4	2.8 ± 0.8^a	3.3 ± 0.8^a	3.3 ± 0.7^a	3.3 ± 0.7^a	3.2 ± 0.8^b
	6	3.7 ± 1.0^a	3.8 ± 1.2^a	4.0 ± 0.9^a	3.9 ± 0.9^a	3.2 ± 1.0^b

Note: Market fried *selroti* is used as control (score 3), with score 1 = bad/hard and score 5 = excellent/soft. Data represents the means scores ± SD of 10 judges. Values bearing different superscripts in each column show statistical difference ($p < .05$).

A, *Leuconostoc mesenteroides* BS1:B1; **B**, all strains of LAB (*Enterococcus faecium* BS1:B2; *Lactobacillus curvatus* BP:B1; *Leuconostoc mesenteroides* BS1:B1; *Pediococcus pentosaceus* BG2:B2); **C**, *Saccharomyces cerevisiae* BA1:Y2; **D**, all strains of yeasts (*Debaryomyces hansenii* BR1:Y4; *Pichia burtonii* BG1:Y1; *Saccharomyces cerevisiae* BA1:Y2; *S. kluyveri* S3:Y3; *Zygosaccharomyces rouxii* S1:Y6); **E**, mixture of B and D (LAB + yeasts, respectively); **F**, mixture of *Leuconostoc mesenteroides* BS1:B1 and *Saccharomyces cerevisiae* BA1:Y2; **G**, without inoculum.

Source: Data from Yonzan, 2007.

(acidifying ability, antagonistic properties), and higher enzymatic profiles than most of the other genera. *Saccharomyces cerevisiae* BA1:Y2 was selected based on its vigorous fermentative property and a wide spectrum of enzymes. Yeast contributed in flavor development to a fermented maize product where LAB were responsible for acidification (Nche 1995). None of the other starters containing strain combinations of *E. faecium*, *Lb. curvatus*, *P. pentosaceus*, *D. hansenii*, *P. burtonii*, *S. kluyveri*, and *Z. rouxii* could produce an organoleptically acceptable *selroti* product (Table 5.2). The results of a consumers' preference trial showed that *selroti* batter prepared by a cell-suspension mixture of *Leuc. mesenteroides* BS1:B1 and *S. cerevisiae* BA1:Y2 was more acceptable than *selroti* batters prepared by the conventional method. This fried *selroti* had a desirable sweet taste, a typical *selroti* flavor, and a soft texture with a golden brown color. Local consumers generally prefer a fried *selroti* with a soft texture, a sweet taste, and a golden brown color.

The principal requirements of the strains are rapid production of CO_2 from maltose and glucose, and generation of good bread flavors (Decock and Cappelle 2005), which are performed by both isolates (*Leuc. mesenteroides* and *S. cerevisiae*) in *selroti* batters. Application of starter cultures may appear appropriate in *selroti* batter production at the household level, since it is cost effective and may contribute to effective control and safeguarding of the fermentation process. Commercial starter cultures of the yeast-bacterial combinations are now available for sourdough production (Decock and Cappelle 2005) (Figure 5.11).

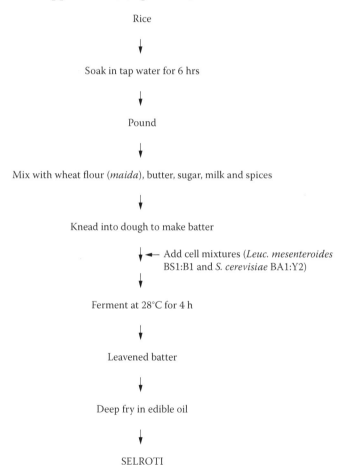

Rice

↓

Soak in tap water for 6 hrs

↓

Pound

↓

Mix with wheat flour (*maida*), butter, sugar, milk and spices

↓

Knead into dough to make batter

↓ ←— Add cell mixtures (*Leuc. mesenteroides* BS1:B1 and *S. cerevisiae* BA1:Y2)

↓

Ferment at 28°C for 4 h

↓

Leavened batter

↓

Deep fry in edible oil

↓

SELROTI

Figure 5.11 Flow sheet of improved method of *selroti* preparation using pure culture combinations.

5.2.2   Jalebi

Lactobacillus fermentum, Lb. buchneri, Lb. bulgaricus, Streptococcus thermo-philus, Strep. lactis, Enterococcus faecalis, Saccharomyces bayanus, S. cerevisiae, and *Hansenula anomala* have been isolated from fermented batter of *jalebi* (Ramakrishnan 1979; Batra 1981, 1986). During fermentation of *jalebi* at 28°C, the bacterial and yeast counts increased from 3.3×10^5 to 12.6×10^6 cfu/g and from 9.4×10^4 to 6×10^6 cfu/g, respectively. At 19°C, while the bacterial count was lowered to 10^6 cfu/g, there was no change in the count of yeasts. During fermentation, the pH decreased from 4.4 to 3.3, and the volume of the batter increased by 9% (Batra 1986).

Detailed microbiological, biochemical, and nutritional studies of other Himalayan fermented cereals except *selroti* and *jalebi* are scarce. I could not collect any publications or literature on the microbiology and biochemistry of *nan,* even though it is a popular wheat flour (*maida*)-based (baked or tandoori) item in Indian cuisine.

5.3   Nutritive value

The nutritional composition of *selroti* is presented in Table 5.3 (Yonzan 2007). Moisture content in *selroti* batters is higher than that of the raw materials due to soaking prior to fermentation and to the addition of water and milk during its preparation. There was a remarkable increase in water-soluble and TCA-soluble nitrogen in *selroti* batters due to solubilization of proteins, indicating the high protein digestibility of the batters (Yonzan 2007). The food value of fermented batters was found to be increased slightly over that of the unfermented raw materials. *Selroti* also had a comparatively higher mineral content (Yonzan 2007). Fermentation may have enhanced the nutritional and mineral contents of the cereals (Amoa and Muller 1973; Blandino et al. 2003; Umeta et al. 2005). The proximate and final food values of *selroti* batters were almost the same as reported in other fermented cereal foods such as *idli* (Soni and Sandhu 1989a, 1989b), *rabadi* (Gupta et al. 1992b), and *tarhana* (Erbas et al. 2005a, 2005b). However, the amino nitrogen and free sugar contents decreased during fermentation of *jalebi* (Ramakrishnan 1979).

Jalebi contains 32%–38% moisture and, per 100 g dry matter: 4–7 g protein, 15–20 g fat, 75–78 g carbohydrate, 2–4 g fiber, 2–3 g ash, 460–480 kcal food value, 2 mg Na, 90 mg K, 70 mg Ca, 10 mg Mg, 1 mg Fe, 0.1 mg Cu, 0.5 mg Zn, 0.17 mg thiamine, 0.03 mg riboflavin, 2.0 mg niacin, 14 mg folic acid, retinol, carotene, and vitamins C and D (Campbell-Platt 1987).

The flour obtained from medium hard wheat, having an ash content of 0.5%–0.6% and sedimentation value of 20–28 ml, was found to be desirable for the preparation of *nan* (Haridas Rao 2000).

Table 5.3 Nutritional Composition of Raw Materials
and Fermented Product (*Selroti* batter)

Parameter[a]	Raw materials		*Selroti* batter	
	Rice	Wheat Flour	Sikkim	Darjeeling hills
Moisture (%)	16.3 ± 0.4	18.4 ± 0.7	42.5 ± 4.8	41.4 ± 4.4
pH	5.5 ± 0.1	5.9 ± 0.1	5.8 ± 0.1	5.7 ± 0.6
Acidity (% as lactic acid)	0.09 ± 0.01	0.1 ± 0.01	0.08 ± 0.01	0.10 ± 0.01
Ash (% DM)	0.7 ± 0.06	0.5 ± 0.07	0.7 ± 0.02	0.8 ± 0.1
Reducing sugar (%)	0.01 ± 0.01	0.02 ± 0.01	2.1 ± 0.5	2.0 ± 0.9
Total sugar (%)	63.8 ± 0.9	58.4 ± 1.2	69.7 ± 5.8	68.3 ± 4.7
Fat (% DM)	1.0 ± 0.01	0.9 ± 0.01	2.8 ± 0.2	2.6 ± 0.1
Water-soluble nitrogen (% DM)	0.016 ± 0.01	0.056 ± 0.01	0.87 ± 0.02	0.048 ± 0.03
TCA-soluble nitrogen (% DM)	0.0016 ± 0.001	0.0017 ± 0.003	0.005 ± 0.004	0.003 ± 0.001
Protein (% DM)	8.3 ± 0.01	11.0 ± 0.5	5.8 ± 1.0	5.9 ± 0.3
Carbohydrate (% DM)	90.0 ± 1.0	87.6 ± 0.9	90.7 ± 1.2	91.5 ± 0.5
Food value (kcal/100 g DM)	42.2 ± 0.4	402.5 ± 0.5	410.8 ± 0.8	409.8 ± 0.4
Sodium (mg/100 g)	5.9 ± 0.7	5.9 ± 0.5	8.8 ± 1.1	8.9 ± 0.3
Potassium (mg/100 g)	47.4 ± 1.1	117.5 ± 2.5	36.9 ± 0.4	24.2 ± 0.5
Calcium (mg/100 g)	9.4 ± 0.5	20.8 ± 0.2	28.8 ± 1.0	24.2 ± 4.4

[a] DM, dry matter; TCA, trichloroacetic acid.
Source: Data from Yonzan, 2007.

chapter six

Ethnic fish products

Fish and fish products in the local Himalayan diet are consumed at a comparatively lower level than other fermented products such as vegetables, milk, and meat products (Thapa 2002). This may be attributed to the pastoral system of agriculture and the preference for consumption of meat and milk products in the region. Societies that are purely pastoral typically lack the custom of eating fish (Ishige 1993). The people of the Himalayas—mostly the eastern Himalayan regions of Nepal, Darjeeling hills, Sikkim, Assam, Arunachal Pradesh, Meghalaya, Manipur, Nagaland, Mizoram, and Tripura, and some regions in Bhutan—consume different types of traditionally processed smoked, sun- or air-dried, fermented, or salted ethnic fish products. Some of these products are prepared using the indigenous knowledge of the rural people on methods of fish preservation. Such fish-processing techniques are still practiced in those regions or villages that are located near water bodies with plenty of freshwater fish. Some villagers sell fish and fish products in the local markets. As the products are manufactured by the rural people during appropriate season, they are regarded as a special dish. The fish products also seem to be an important source of protein in the local diet.

According to some village elders in these regions, interviewed during the survey, *suka ko maacha* was produced in bulk in most of the places in the eastern part of Nepal located near river sites—even in the low altitudes of the Balasan, Teesta, and Rangit rivers in Darjeeling hills and Sikkim—until the 1960s (Thapa 2002). Nowadays, production of these traditional processed fish products is confined to only a few areas and is hardly seen in the local markets due to declines in the fish population in hill rivers. The study reveals that the decline in the fish population in the Teesta and Rangit Rivers in Darjeeling–Sikkim is mainly due to hydropower project activities in the hills, leading to soil erosion and siltation, and to water pollution caused by the growth of industry, sewage, and increased use of pesticides (Tamang 2005c).

The traditional technique for fish preservation in the Eastern Himalayan regions involves dehydration (drying), smoking, fermentation, and salting (low-salt) for preservation (Thapa et al. 2004). Dehydration, smoking, salting, and fermentation are the best methods for preservation of available perishable fish (Beddows 1985). The products are whole fish obtained from freshwater lakes, and these are eaten as a side dish, a curry, or a pickle. Traditionally, no fish sauce or shrimp products are prepared and used as a condiment or seasoning in the local diet in the Eastern Himalayan regions. This may be due to the use of ginger, onion, and other spices to stimulate the appetite instead of using a umami-taste producer such as fish sauce and soy sauce.

6.1   Important ethnic fish products

Gnuchi is an ethnic smoked and dried fish product of the Lepcha of Darjeeling hills and Sikkim. *Suka ko maacha, sidra,* and *sukuti* are sun-dried fish products common in the diet of the ethnic people of Nepal, Darjeeling hills, Sikkim, and Bhutan as a side dish or pickle. *Ngari* and *hentak* are a unique fermented fish cuisine of the Meitei in Manipur. *Tungtap* is an ethnic fermented fish product of the Khasi in Meghalaya. *Karati, bordia,* and *lashim* are sun-dried and salted fish products commonly sold in the local markets in Assam (Thapa 2002).

6.1.1   Suka ko maacha

Suka ko maacha is an ethnic smoked-fish product commonly prepared in Nepal, Darjeeling hills, Sikkim, and the southeastern part of Bhutan. *Suka* (dry) *maacha* (fish) means "dry fish" in the Nepali language.

6.1.1.1   Indigenous knowledge of preparation

The hill river fishes, mostly *dothay asala* (*Schizothorax richardsonii* Gray) and *chuchay asala* (*Schizothorax progastus* McClelland), are caught and collected in a bamboo basket, locally called a *bhukh*, from rivers or streams. Fish are degutted, washed, and mixed with salt and turmeric powder. Degutted fish are hooked on a bamboo string and hung above an earthen oven in the kitchen for 7–10 days; the smoked fish is called *suka ko maacha* (Figure 6.1). It can be kept up to 6 months at room temperature for consumption. *Suka ko maacha* is kept inside a *perungo,* a closed basket of bamboo, and sold in the local markets at Aitabare, Therathum, Dhankuta, Elam, and Dharan in the eastern districts of Nepal (Figure 6.2).

6.1.1.2   Culinary practices

Suka ko maacha is prepared as a curry by mixing with tomato, chilli, and salt. It is also cooked with vegetables and is eaten as a side dish with boiled rice.

6.1.1.3   Microorganisms

Lactic acid bacteria (LAB): *Lactococcus lactis* subsp. *cremoris, Lc. lactis* subsp. *lactis, Lc. plantarum, Leuconostoc mesenteroides, Enterococcus faecium, E. faecalis, Pediococcus pentosaceus;* yeasts: *Candida chiropterorum, C. bombicola,* and *Saccharomycopsis* spp. (Thapa et al. 2006).

Fish

↓

Catch from river

↓

Gut and wash

↓

Mix with turmeric powder and salt

↓

Hook in a bamboo string or spread on a bamboo tray

↓

Smoke above the earthen-oven for 7–10 days

↓

SUKA KO MAACHA

Figure 6.1 Indigenous method of *suka ko maacha* preparation in Nepal.

Figure 6.2 Suka ko maacha of Nepal.

6.1.2   Gnuchi

Gnuchi is a typical smoked and dried fish product common to the Lepcha tribes of Darjeeling hills of Kalimpong subdivision living near the Ghish khola and Teesta rivers in Sikkim. The word *gnuchi* means smoked fish in the Lepcha.

6.1.2.1   Indigenous knowledge of preparation

Fish are collected in a bamboo basket, locally called a *tamfyok*, which is woven of bamboo strips and is properly tied around the waist of the fisherman while fishing. Fish that are captured include *Schizothorax richardsonii* Gray, *Labeo dero* Hamilton, *Acrossocheilus* spp., *Channa* sp., etc. According to the Lepcha fishermen, the best variety for *gnuchi* is *Schizothorax* sp. Fish are placed on a large bamboo tray, locally called a *sarhang*, to drain off water; they are then degutted and mixed with salt and turmeric powder. Fish are separated according to their size. The larger fish are spread in an upside-down manner on a *sarhang* and kept above an earthen oven in the kitchen. The small fish are hung one after the other on a bamboo strip above the earthen oven and kept for 10–14 days, when the *gnuchi* is ready to be consumed (Figure 6.3). *Gnuchi* can be kept at room temperature for 2–3 months.

6.1.2.2   Culinary practices

Gnuchi is eaten as a curry with boiled rice. It is also cooked with vegetable.

6.1.2.3   Microorganisms

Lactic acid bacteria (LAB): *Lactococcus lactis* subsp. *cremoris*, *Lc. lactis* subsp. *lactis*, *Lc. plantarum*, *Leuconostoc mesenteroides*, *Enterococcus faecium*, *E. faecalis*, *Pediococcus pentosaceus*; yeasts: *Candida chiropterorum*, *C. bombicola*, *Saccharomycopsis* spp. (Thapa et al. 2006).

6.1.3   Sidra

Sidra is a sun-dried fish product commonly consumed by the Nepali of the Himalayas. *Sidra* and *sukuti* are one of the most indispensable ethnic food items of the Nepali (Figure 6.4).

6.1.3.1   Indigenous knowledge of preparation

During its preparation, the whole fish (*Puntius sarana* Hamilton) is collected, washed, and dried in the sun for 4–7 days (Figure 6.5). *Sidra* can be stored at room temperature for 3–4 months for consumption.

6.1.3.2   Culinary practices

Sidra is consumed as a pickle. During pickle making, *sidra* is roasted and mixed with dry chilli, boiled tomato, and salt to make a thick pickle paste.

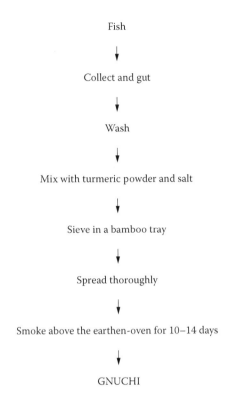

Fish

↓

Collect and gut

↓

Wash

↓

Mix with turmeric powder and salt

↓

Sieve in a bamboo tray

↓

Spread thoroughly

↓

Smoke above the earthen-oven for 10–14 days

↓

GNUCHI

Figure 6.3 Indigenous method of *gnuchi* preparation in Kalimpong.

In the typical meal of the Nepali, cooked rice and *khalo dal* (black gram soup) is served with *sidra ko acchar* or pickle. It is sold at local market in these regions.

6.1.3.3   Microorganisms

Lactic acid bacteria: *Lactococcus lactis* subsp. *cremoris, Lc. lactis* subsp. *lactis, Lc. plantarum, Leuconostoc mesenteroides, Enterococcus faecium, E. faecalis, Pediococcus pentosaceus,* and *Weissella confuse;* yeasts: *Candida chiropterorum, C. bombicola,* and *Saccharomycopsis* spp. (Thapa et al. 2006).

6.1.4   Sukuti

Sukuti is a very popular sun-dried fish product in Nepal, Darjeeling hills, Sikkim, and Bhutan (Figure 6.6).

6.1.4.1   Indigenous knowledge of preparation

During preparation of *sukuti*, the fish (*Harpodon nehereus* Hamilton) are collected, washed, rubbed with salt, and dried in the sun for 4–7 days

Figure 6.4 *Sidra* of Darjeeling hills.

Fish

↓

Wash

↓

Sun-dry for 4–7 days

↓

SIDRA

Figure 6.5 Indigenous method of *sidra* preparation in Darjeeling hills.

(Figure 6.7). The sun-dried fish product can be stored at room temperature for 3–4 months for consumption.

6.1.4.2   Culinary practices

Sukuti is consumed as pickle, soup, and curry (Figure 6.8). During curry preparation, *sukuti* is fried and mixed with dry chilli, onion, and salt to make a pickle. It is usually eaten with boiled rice and black gram soup. It is also commonly sold at local markets.

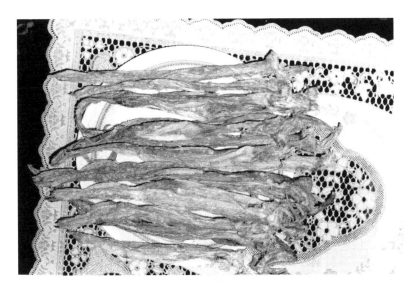

Figure 6.6 *Sukuti* of Darjeeling hills.

Fish

↓

Wash

↓

Rub with salt

↓

Sun dry for 4–7 days

↓

SUKUTI

Figure 6.7 Indigenous method of *sukuti* preparation in Sikkim.

6.1.4.3 Microorganisms

Lactic acid bacteria (LAB): *Lactococcus lactis* subsp. *cremoris*, *Lc. lactis* subsp. *lactis*, *Lc. plantarum*, *Leuconostoc mesenteroides*, *Enterococcus faecium*, *E. faecalis*, *Pediococcus pentosaceus*; yeasts: *Candida chiropterorum*, *C. bombicola*, and *Saccharomycopsis* spp. (Thapa et al. 2006).

Figure 6.8 Traditional pickles made from *sidra* and *sukuti*.

6.1.5   Ngari

Ngari is a traditional fermented fish product of Manipur. *Ngari* preparation and consumption reflects the typical food culture of Manipur.

6.1.5.1   Indigenous knowledge of preparation

During traditional preparation of *ngari*, the whole fish, locally called *phoubu* (*Puntius sophore* Hamilton), is collected from a river, rubbed with salt, and dried in the sun for 3–4 days. Then, the sun-dried fish are washed briefly and spread on bamboo mats and then pressed tightly in an earthen pot (Figure 6.9). A layer of mustard oil is applied to the inner wall of the pot before adding the fish. The pot is sealed airtight and then stored at room temperature for 4–6 months (Figure 6.10). After fermentation, the lid is opened, and the *ngari* is ready for consumption. It can be kept for more than a year at room temperature.

6.1.5.2   Culinary practices and economy

Ngari is one of the most delicious food items in the diet of the Manipuri people. It is eaten daily as a popular side dish called *ironba* (*ngari* mixed with potato, chillies, etc.) with cooked rice. Sometimes, it is mixed

Figure 6.9 *Ngari* in a narrow-mouthed big earthen pot in Manipur.

with meat and vegetable and is eaten with boiled rice. *Ngari* is sold in narrow-mouthed big earthen pots in all local markets in Manipur (Figure 6.11).

6.1.5.3 Microorganisms

Lactic acid bacteria (LAB): *Lactococcus lactis* subsp. *cremoris, Lc. plantarum, Enterococcus faecium, Lactobacillus fructosus, Lb. amylophilus, Lb. coryniformis* subsp. *torquens,* and *Lb. plantarum*; bacilli: *Bacillus subtilis* and *B. pumilus*; aerobic cocci: *Micrococcus* spp.; yeasts: *Candida* and *Saccharomycopsis* spp. (Thapa et al. 2004).

6.1.6 Hentak

Hentak is a ball-like thick paste prepared by fermentation of a mixture of sun-dried fish powder and petioles of aroid plants in Manipur.

6.1.6.1 Indigenous knowledge of preparation

This finger-sized fish (*Esomus danricus* Hamilton) is collected and washed thoroughly and then sun dried. The dried fish are crushed to a powder. Petioles of *Alocasia macrorhiza* are cut into pieces, washed with water, and

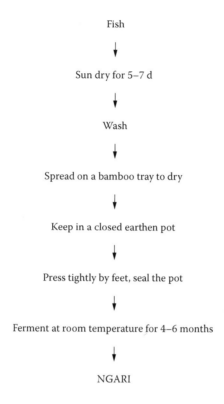

Fish

↓

Sun dry for 5–7 d

↓

Wash

↓

Spread on a bamboo tray to dry

↓

Keep in a closed earthen pot

↓

Press tightly by feet, seal the pot

↓

Ferment at room temperature for 4–6 months

↓

NGARI

Figure 6.10 Indigenous method of *ngari* preparation in Manipur.

then exposed to sunlight for one day. An equal amount of the cut pieces of the petioles of *Alocasia macrorhiza* is mixed with powdered fish, and a ball-like thick paste is made. The ball-like mixture is kept in an earthen pot, and the opening of the earthen pot is tightly sealed by a cloth or a lid to maintain anaerobic conditions inside the container, and is fermented for 7–9 days, preferably under sunshine (Figure 6.12). *Hentak* is prepared at every house in Manipur.

6.1.6.2   Culinary practices

Hentak is consumed as a curry as well as a condiment with boiled rice. Sometimes, *hentak* is given in small amounts to mothers in confinement and patients in convalescence. *Hentak* is one of the most delicious indigenous fish products in the local diet of Manipur. It is not sold in the local market.

6.1.6.3   Microorganisms

Lactic acid bacteria (LAB): *Lactococcus lactis* subsp. *cremoris, Lc. plantarum, Enterococcus faecium, Lactobacillus fructosus, Lb. amylophilus, Lb. cornifomis*

Figure 6.11 *Ngari* is sold at Imphal by Meitei women.

subsp. *torquens*, and *Lb. plantarum*; endospore rods: *Bacillus subtilis* and *B. pumilus*; aerobic cocci: *Micrococcus* spp.; yeasts: *Candida* and *Saccharomycopsis* spp. (Thapa et al. 2004).

6.1.7   Tungtap

Tungtap is a fermented fish paste, commonly consumed by the Khasi and Garo of Meghalaya. *Tungtap* is one of the most delicious side dishes in the cuisine of Meghalaya.

6.1.7.1   Indigenous knowledge of preparation
During preparation of *tungtap*, sun-dried fish (*Danio* spp.) are collected, washed briefly, and mixed with salt. The salted, sun-dried fish are kept in an airtight earthen pot and fermented for 4–7 days (Figure 6.13), after which they are ready for consumption (Figure 6.14).

6.1.7.2   Culinary practices
Tungtap is consumed as a pickle, mixed with onion, dry chilli, garlic, and seeds of *Zanthoxylum nitidum*. Sometimes, it is cooked to make curry and is eaten with rice.

Figure 6.12 Indigenous method of *hentak* preparation in Manipur.

Figure 6.13 Indigenous method of *tungtap* preparation in Meghalaya.

6.1.7.3   Microorganisms

Lactic acid bacteria (LAB): *Lactococcus lactis* subsp. *cremoris*, *Lc. plantarum*, *Enterococcus faecium*, *Lactobacillus fructosus*, *Lb. amylophilus*, *Lb. corynifomis* subsp. *torquens*, and *Lb. plantarum*; endospore rods: *Bacillus subtilis* and *B. pumilus*; aerobic cocci: *Micrococcus* spp.; yeasts: *Candida* and *Saccharomycopsis* spp. (Thapa et al. 2004).

6.1.8   Dried fishes of Assam

Karati, *bordia*, and *lashim* are similar types of sun-dried and salted fish products sold at Shillong and Guwahati markets (Figure 6.15).

Figure 6.14 *Tungtap* of Meghalaya.

Figure 6.15 Ethnic fish products sold at Shillong by Khasi woman.

Fish

↓

Collect

↓

Wash and mix with salt

↓

Dry in the sun for 5–7 days

↓

KARATI/ BORDIA/LASHIM

Figure 6.16 Indigenous method of preparation of some sun-dried and salted fish products in Assam.

6.1.8.1   Indigenous knowledge of preparation

During preparation of dried-fish products, fish (*Gudusia chapra, Pseudeutropius atherinoides,* and *Cirrhinus reba*) are collected, washed, and rubbed with salt and dried in the sun for 4–7 days. The sun-dried fish products are stored at room temperature for 3–4 months for consumption (Figure 6.16). *Karati* is prepared from *Gudusia chapra* Hamilton, *bordia* is prepared from *Pseudeutropius atherinoides* Bloch, and *lashim* is prepared from *Cirrhinus reba* Hamilton.

6.1.8.2   Culinary practices

These fish products are eaten as a side dish along with boiled rice in Assam and other northeastern parts of India.

6.1.8.3   Microorganisms

Lactic acid bacteria (LAB): *Lactococcus lactis* subsp. *cremoris, Leuconostoc mesenteroides,* and *Lactobacillus plantarum*; bacilli: *Bacillus subtilis* and *B. pumilus*; yeast: *Candida* spp. (Thapa et al. 2007).

6.2   Microbiology

Microbial analysis of traditionally processed fish products of the Himalayas shows that the colony-forming units (cfu) per gram of samples is 10^8, indicating the predominance of microbes (see Tables 6.1–6.3). The population of *Bacillus* species was below 10^4 cfu/g. However, no species of *Bacillus* were recovered from a few samples of *suka ko maacha* and *sukuti* (Thapa et al. 2006). Yeasts were recovered from only a few samples, such

Table 6.1 Microbial Population of Traditionally Processed Fish Products
of Nepal, Darjeeling Hills, and Sikkim

Product	Log cfu/g dry weight [a]			
	Lactic acid bacteria	Bacterial endospores	Yeast	Aerobic mesophilic count
Suka ko maacha	7.3 (5.4–8.3)	2.7 (0–4.6)	3.2 (2.5–4.0)	7.6 (6.1–8.5)
Gnuchi	6.6 (5.8–6.9)	3.2 (3.1–3.5)	0	6.8 (6.3–7.0)
Sidra	5.3 (4.8–6.5)	3.1 (2.8–3.3)	0	5.8 (5.2–6.8)
Sukuti	5.4 (4.7–6.2)	0	0	5.7 (5.1–6.8)

Note: Data represent the mean values, and ranges are given in parentheses. Molds were not
detected.

[a] Microbiological data are transformed into logarithms of the numbers of colony-forming
units (cfu/g).

Table 6.2 Microbial Population of Fermented Fish Products
of Manipur and Meghalaya

Product	Log cfu/g dry weight [a]			
	Lactic acid bacteria	Bacterial endospores	Yeast	Aerobic mesophilic count
Ngari	6.8 (5.8–7.2)	4.2 (3.3–4.6)	3.1 (2.8–3.3)	7.0 (6.3–7.2)
Hentak	4.6 (4.0–4.8)	3.8 (3.3–4.2)	0	4.7 (4.3–4.9)
Tungtap	5.9 (5.2–6.2)	3.2 (3.5–3.7)	3.0 (2.6–3.5)	6.2 (5.9–6.4)

Note: Data represent the mean values, and ranges are given in parentheses. Molds are not
detected.

[a] Microbiological data are transformed into logarithms of the numbers of colony-forming
units (cfu/g).

Table 6.3 Microbial Population of Sun-Dried Fish Products of Assam

Microorganisms	Log cfu/g dry weight [a]		
	Karati	*Bordia*	*Lashim*
Lactic acid bacteria	4.2 (4.0–4.4)	5.3 (4.3–5.6)	5.8 (4.2–6.2)
Bacterial endospores	3.1 (2.8–3.3)	3.5 (2.9–3.8)	2.1 (1.4–2.6)
Yeast	3.1 (2.7–3.3)	2.2 (1.8–2.5)	3.0 (2.2–3.3)
Aerobic mesophilic count	5.1 (4.2–5.5)	5.6 (5.2–6.0)	6.0 (5.6–6.3)

Note: Data represent the mean values, and ranges are given in parentheses. Molds were not
detected.

[a] Microbiological data are transformed into logarithms of the numbers of colony-forming
units (cfu/g).

as *suka ko maacha, ngari, tungtap, karati,* and *bordia,* at the level of 10^3 cfu/g (Thapa 2002). About 70% of LAB, 15% of bacilli, 5% micrococci, and 10% yeasts have been reported in the fish products of the Himalayas (Thapa 2002). Bacterial fermentation is more dominant than yeast and mold fermentation in fish.

Coccus lactic acids are identified as *Lactococcus lactis* subsp. *cremoris* Schleifer et al., *Lc. plantarum* Schleifer et al., *Lc. lactis* subsp. *lactis* Schleifer et al., *Leuconostoc mesenteroides* (Tsenkovskii) van Tieghem, *Enterococcus faecium* (Oral-Jensen) Schleifer and Kilpper-Bälz, and *E. faecalis* (Andrewes and Horder) Schleifer and Kilpper-Bälz. Tetrads are identified as *Pediococcus pentosaceus* Mees. Heterofermentative LAB are identified as *Lactobacillus confusus* (Holzapfel and Kandler) Sharpe, Garvie, and Tilbury; and *Lb. fructosus* Kodama. Homofermentative lactic acids are identified as *Lb. amylophilus* Nakamura and Crowell, *Lb. coryniformis* subsp. *torquens* Abo-Elnaga and Kandler, and *Lb. plantarum* Orla-Jensen. Hetero- and homofermentative acids were identified on the basis of sugar fermentation using the API system, lactic acid isomer, and *meso*-diaminopimelic acid determination (Thapa et al. 2004, 2006).

It has been observed that, among LAB, cocci dominate the lactic microflora in the Himalayan fish products (Thapa 2002). This may be due to gradations of concentration of salts used during processing, which control the bacterial flora (Tanasupawat et al. 1993). None of the LAB strains obtained from these samples are halotolerant (i.e., 18% salt tolerance). LAB species are also reported from other Asian fish products, such as species of *Lactobacillus* and *Pediococcus* from *nam-pla* and *kapi*, fermented fish products of Thailand (Watanaputi et al. 1983; Tanasupawat et al. 1992), and *Pediococcus acidilactici* and *Leuconostoc paramesenteroides* from *burong isda*, a fermented-rice–freshwater-fish mixture of the Philippines (Mabesa et al. 1983).

Endospore-forming rods are identified as *B. subtilis* (Ehrenberg) Cohn and *B. pumilus* Meyer and Gottheil. Though the load of bacilli is around 10^4 cfu/g, their presence shows the dominance in fish products next to LAB. *Bacillus* species are found to be predominant in the fish products due to their ability as endospore formers to survive under the prevailing conditions (Crisan and Sands 1975). Many workers have reported the presence of *Bacillus* species in several traditionally processed fish products such as in *nam-pla* and *kapi* of Thailand (Watanaputi et al. 1983) and *nga-pi* of Myanmar (Tyn 1993). Species of *Micrococcus* are also present in the Himalayan fish products (Thapa 2002). Species of *Micrococcus* have also been reported from some fermented fish products of Thailand (Phithakpol 1987) and *shiokara* of Japan (Fujii et al. 1994; Wu et al. 2000). Two genera of yeasts, *Candida* and *Saccharomycopsis* species, have been recovered from some samples of the Himalayan fish products (Thapa 2002).

6.2.1   Pathogenic contaminants

Bacillus cereus is present in 66% of total samples of the Himalayan fish products (Thapa 2002). However, none of the samples was found to contain more than 10^2 cfu/g of *B. cereus* population. The low number of *B. cereus* in foods is not considered significant (Roberts et al. 1996). Initial growth of *Bacillus cereus* is observed in fish sauce but is inhibited gradually due to growth of LAB (Aryanta et al. 1991). The load of *Staphylococcus aureus* was also found to be less than 10^3 cfu/g in all samples tested. *Staphylococcus aureus* is regarded as a poor competitor, and its growth in fermented foods is generally associated with a failure of the normal microflora (Nychas and Arkoudelos 1990). *Staphylococcus aureus* survives during *shiokara* fermentation but does not produce enterotoxin, confirming the safety of traditional *shiokara* of Japan (Wu et al. 1999). Enterobacteriaceae occurs widely in all samples of the Himalayan fish product samples, however, the population is less than 10^3 cfu/g (Thapa 2002). The presence of *Bacillus cereus, Staphylococcus aureus*, and Enterobacteriaceae in fish products may be due to contamination during processing either through smoking or drying. However, the population of these contaminants is not more than 10^3 cfu/g in the fish product samples tested, which would be the impact of competition and/or antagonistic reaction of predominant LAB that have prevented their proliferation. All of these fermented and sun-dried/smoked fish are taken daily or more frequently in the meals of the people of the Eastern Himalayas, and there have been no reports or documentation of food poisoning or toxicity. It may be claimed that, despite the presence of some pathogenic contaminants in the finished products, the large and dominant populations of LAB have helped to make these traditionally processed fish products safe for consumption by the ethnic people for centuries. Salting in the initial stages of traditional fish processing can also inhibit the growth of pathogenic microorganisms.

6.2.2   Functional or technological properties

Microorganisms isolated from the Himalayan fish products have enzymatic activities during processing of the fish products. Three strains, viz., *Enterococcus faecium* GG6, *Lactobacillus coryniformis* subsp. *torquens* T2:L1, and *Leuconostoc mesenteroides* BA4, isolated from *gnuchi, tungtap, bordia*, respectively, show proteolytic activity with low protease activity (1.0 U/ml). In contrast, all *Bacillus* strains show proteolytic activity with protease activity of 4 U/ml (Thapa 2002; Thapa et al. 2006). This indicates that LAB have very low proteolytic activities in the fish products. Some strains of LAB but all strains of *Bacillus subtilis* show amylolytic activity, with 3.2 U/ml to 5.8 U/ml α-amylase activity. LAB as well as *Bacillus subtilis* with amylolytic activities are essential in liquefaction during processing of fish products. Proteolysis and liquefaction that occur during fish production have been reported to be largely the result of autolytic breakdown of the fish tissues, which is more rapid when whole fish are used, since the head and viscera contain higher concentrations of proteolytic enzymes than other tissues (Backhoff 1976). Some strains of LAB and *Bacillus* show lipolytic activity on tributyrin agar plates. Activities of peptidase and esterase-lipase (C4 and C8) shown by strains of LAB may help in developing the acceptable flavors in Himalayan fish products.

LAB strains from Himalayan fish products show antagonistic properties against many pathogenic bacteria (Thapa 2002). However, none of the LAB strains are found to produce bacteriocin. Although there is a report of biogenic amines in many fish products (Halász et al. 1994), and LAB frequently produce histamine and tyramine in processed fish (Leisner et al. 1994), none of the LAB strains are found to decarboxylase the used amino acids—tyrosine, lysine, histidine, and ornithine—in the Himalayan fish products (Thapa et al. 2006). These traditionally processed fish products are safe to eat. However, the lack of histamine, tyramine, cadaverine, and putrescine producers isolated from traditionally processed fish products in the study (Thapa 2002) could possibly be explained by the lack of free amino acids within the samples. The concentration of amino acids in food is important for biogenic amine formation (Joosten and Northolt 1989). Before confirming the nonproduction of biogenic amine in the traditionally processed fish products of the Himalayas, qualitative and quantitative analysis of biogenic amines is necessary.

Some strains of LAB isolated from the Himalayan fish products show high degrees of hydrophobicity (>75%), among which *P. pentosaceus* GG2 (isolated from *gnuchi*) shows the highest degree of hydrophobicity of 94%, indicating strong hydrophobic properties (Thapa et al. 2006). All strains of LAB have more than 30% hydrophobicity, indicating their adherence abilities. The adherence of microorganisms to various surfaces seemed to be mediated by hydrophobic interactions (Rosenberg 1984). A high degree of hydrophobicity by some strains of LAB from the lesser-known Himalayan fish products probably indicates a probiotic character, since hydrophobicity is one of the important factors in determining the probiotic property of the microorganisms (Tamang and Holzapfel 2004).

6.3   Nutritive value

The nutritional composition of traditionally processed fish products *suka ko maacha, gnuchi, sidra, sukuti, ngari, hentak, tungtap, karati, bordia*, and *lashim* as well as fresh river fish of Sikkim is presented in Table 6.4. All of these fish products are acidic in nature, with the pH and acidity ranging between 6.2 to 6.5 and 0.5% to 1.1%, respectively, due to predominance of lactic acid

Table 6.4 Nutriional Composition of Ethnic Fish Products of the Eastern Himalayas

Product	pH	Percent (%)					Food value (kcal/100 g)	Minerals (mg/100 g)				
		Moisture	Ash	Protein	Fat	Carbohydrate		Ca	Fe	Mg	Mn	Zn
Suka ko maacha	6.4	10.4	16.2	35.0	12.0	36.8	395.2	38.7	0.8	5.0	1.0	5.2
Gnuchi	6.3	14.3	16.9	21.3	14.5	47.3	404.9	37.0	1.1	8.8	1.1	7.5
Sidra	6.5	15.3	16.6	25.5	12.2	45.7	394.6	25.8	0.9	1.6	0.8	2.4
Sukuti	6.4	12.7	3.6	36.8	11.4	38.2	402.6	17.7	0.3	1.4	0.2	1.3
Ngari	6.2	33.5	21.1	34.1	13.2	31.6	381.6	41.7	0.9	0.8	0.6	1.7
Hentak	6.5	40.0	5.0	32.7	13.6	38.7	408.0	38.2	1.0	1.1	1.4	3.1
Tungtap	6.2	35.4	8.9	32.0	12.0	37.1	384.4	25.8	0.9	1.6	0.8	2.4
Karati	6.3	11.8	4.5	35.0	12.4	38.1	404.0	ND	ND	ND	ND	ND
Bordia	6.4	12.0	15.3	24.5	12.3	47.9	400.3	ND	ND	ND	ND	ND
Lashim	6.4	9.6	12.8	28.3	11.8	47.1	407.8	ND	ND	ND	ND	ND

Note: Data represent the means of five samples. ND, not determined.

bacteria in the finished products (Thapa et al. 2004, 2006) and also the subsequent fermentation or processing of fish. Moisture content is low, up to 10% in *suka ko maacha*, although *hentak*, the semi-paste product, contained 40% moisture (Thapa and Pal 2007). Drying in the sun or smoking during preservation result in dehydration, which explains why most of the fish products have a low moisture content. Dried fish are produced with a moisture content of 17% to 45% (Adams 1998). Due to its low moisture content and its slightly acidic nature, the shelf life of the product can be prolonged and can be kept for a longer period at room temperature.

A high protein content is observed in all fish products, indicating their importance as a source of protein intake in the local diet (Thapa and Pal 2007). Fermented fish products are generally high in protein and amino compounds (Puwastien et al. 1999). *Tungtap* has high calcium (5040 mg/100 g) and phosphorus (1930 mg/100 g) among the minerals (Agrahar-Murugkar and Subbulaksmi 2006). The traditional process of drying and further fermentation weakens the bones considerably, almost dissolving them in the flesh portion, which may be the reason for the increase in calcium levels in *tungtap*. The chemical composition of fermented fish products, mainly fish sauce and shrimp paste of Southeast Asia, was similar to soy sauce and soybean paste (Mizutani et al. 1992).

Traditionally processed fish products are unique in these regions, and they are important because they supplement the nutritive value of the local diet. They are also a good example of biopreservation of perishable fish of the mountain rivers and lakes in the Himalayan regions.

6.4   Conclusion

Traditional processing of perishable fish by smoking, drying, salting, and fermentation are the principal methods of biopreservation without refrigeration or addition of any synthetic preservative in the Eastern Himalayas. The indigenous knowledge that the ethnic people use in the production of processed fish for consumption is worth documentation. Although the traditionally processed fish products of the Himalayas are lesser known, the role of LAB in fermentation and in enhancing functional properties—such as a wide spectrum of enzymatic activities as well as enzymatic profiles, antagonistic activities, probiotic properties, and nonproduction of biogenic amine—is a remarkable observation.

Some of the LAB strains having functional properties may be developed as starters for controlled, optimized production of fish products. Diversity within species of LAB may enrich the microbial resources from food ecosystems of the Himalayas. It may be concluded that the fermenting or processing of fish products in the eastern Himalayas, mainly in North East India, is carried out mainly to preserve the perishable fish caught from the rivers, streams, ponds, and lakes for future consumption, rather than as a nutritional supplement.

With the information obtained on the microbiology and nutritive value of fermented or preserved fish products, some of the ethnic fish products of the Himalayas such as *sidra, sukuti, ngari, hentak,* and *tungtap* can be upgraded. Many ethnic people in the eastern Himalayas depend upon the sale of these indigenous fish products as an important part of their economy.

chapter seven

Ethnic meat products

Meat is a part of the daily diet for many ethnic people dwelling in the Himalayan regions of India, Nepal, Bhutan, and Tibet in China. Although chicken, duck, fish, pork, beef, and eggs are popular meats served in the Southeast Asian countries, the serving of these foods is greatly influenced by the religious dictates of the dominant religions (McWilliams 2007). Meat is the flesh (muscle tissue) of warm-blooded animals, and it is highly susceptible to microbial spoilage (Bacus and Brown 1981). Raw meat spoils at high ambient temperatures within a few hours due to its high moisture and protein contents (Dzudie et al. 2003). Consequently, these perishable raw meats are commonly dried or smoked or fermented to prolong their shelf life (Rantsiou and Cocolin 2006).

The domestic livestock of the Himalayas is mostly cattle, sheep, goats, pigs, yaks, etc., which are mainly used for meat, milk, and milk products (Tamang 2005a). Yaks (*Bos grunniens*) are reared mostly on extensive alpine and subalpine scrublands at elevations of 2100–4500 m in the Himalayas for milk products and meat (Sharma et al. 2006).

The Himalayan people have prepared and consumed a variety of traditionally processed smoked, sun-dried, air-dried, or fermented meat products, including ethnic sausages, for centuries. They use their indigenous knowledge of preservation of perishable meats without using starter culture and chemicals. Tibetans, Bhutia, Drukpa, and the Lepcha usually prefer beef, yak, and pork. Beef is taboo to a majority of Nepali, except for the Tamang and Sherpa castes. The Newar caste of Nepali prefers to eat buffalo meat. Naga, Khasi, Garo, Mizo, and several tribes of North East India eat pork and beef. Chicken and chevon (goat meat) are the most popular animal meats for nonvegetarian Himalayan people. However, regular consumption of meat is too expensive for the majority of the poor rural people who live in the Himalayas. They slaughter domestic animals for special occasions, festivals, and marriages. After the ceremony, the fresh animal flesh is cooked and eaten. The remaining meat is preserved by smoking or drying for future consumption. Tibetans, Bhutia, Drukpa, and Lepcha slaughter yaks occasionally, consume the fresh meat, and preserve the rest by smoking or drying or by fermenting into varieties of meat products in the form of smoked meats, preserved animal fats, etc.

The native skills of the Himalayan people in preservation methods of locally available raw meat are apparent in their ability to make sausage-like products using unappealing animal parts such as scraps, organ meats, fat, blood, etc. Traditional sausage-like meat products are also made with the leftover parts of the animal. Hence, *kargyong* (ethnic sausage-like products)-making may be the first of the skills that the high-altitude dwellers used to preserve leftovers meats. Ethnic people might have invented such preservation techniques to feed themselves when food was scarce. The Himalayan ethnic meat products are naturally cured without using starter cultures or addition of sodium nitrites/nitrates.

7.1   Important ethnic meat products

Some of the common as well as lesser-known traditionally processed ethnic meat products of the Himalayas are *kargyong, satchu, suka ko masu, kheuri, chilu, chartayshya, jamma,* and *arjia.*

7.1.1   Lang kargyong

Lang kargyong (Figure 7.1) is an indigenous sausage-like traditional meat product of Sikkim prepared from beef and is mostly consumed by the Bhutia, Tibetans, Drukpa, Lepcha, and Sherpa. It is soft or hard and brownish in color. The Lepcha call it *tiklee.*

Figure 7.1 *Lang* (beef) *kargyong* of Sikkim.

7.1.1.1 Indigenous knowledge of preparation

Lean meat of beef, along with its fat, is chopped finely and combined with crushed garlic, ginger, and salt and then mixed with a small amount of water. No sugar, nitrites, or nitrates are added, unlike with European sausages. The mixture is stuffed into a segment of the gastrointestinal tract, locally called *gyuma*, that is used as a natural casing with a diameter of 3–4 cm and a length of 40–60 cm. One end of the casing is tied with rope, and the other end is sealed after stuffing, after which the casing is boiled for 20–30 min. Cooked sausages are removed and hung on bamboo strips above the kitchen oven for smoking and drying for 10–15 days or more to make *kargyong* (Figure 7.2).

Due to the use of natural casings, *kargyong* has a natural curve shape. The method of preparation of *lang kargyong* is the same for all the ethnic groups, but the ingredients vary from place to place and also from community to community. The shelf life of *kargyong* is one month in high altitudes.

7.1.1.2 Culinary practices and economy

Lang kargyong is eaten after boiling for 10–15 min. It is also sliced and fried in edible oil with onion, tomato, powdered or ground chillies, and salt and is

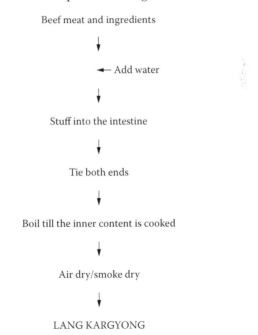

Beef meat and ingredients

↓

◄— Add water

↓

Stuff into the intestine

↓

Tie both ends

↓

Boil till the inner content is cooked

↓

Air dry/smoke dry

↓

LANG KARGYONG

Figure 7.2 Indigenous method of *lang kargyong* preparation in Sikkim.

made into curry. It is also consumed as a fried sausage with *raksi*, a distilled liquor, or *chyang/kodo ko jaanr*, a mildly alcoholic finger-millet-based beverage. *Lang kargyong* may also be eaten as cooked sausage before fermentation.

Lang kargyong is sold in the local restaurants and food stalls in Sikkim, Darjeeling hills, Bhutan, Ladakh, Tibet, etc. It is also prepared for home consumption and also during marriages and festivals. The price of each kg of *kargyong* is Rs. 150–180.

7.1.1.3 Microorganisms
Bacteria: *Lactobacillus sake, Lb. divergens, Lb. carnis, Lb. sanfrancisco, Lb. curvatus, Leuconostoc mesenteroides, Enterococcus faecium, Bacillus subtilis, B. mycoides, B. thuringiensis, Staphylococcus aureus, Micrococcus* spp.; yeasts: *Debaryomyces hansenii* and *Pichia anomala* (Rai 2008).

7.1.2 Yak kargyong

Yak kargyong is an ethnic sausage-like product prepared from yak meat, and is mostly consumed by the Tibetans, Bhutia, Drukpa, Sherpa, and Lepcha of the Himalayas. It is usually prepared during November to December.

7.1.2.1 Indigenous knowledge of preparation
Yak meat and its fat is finely chopped and combined with crushed garlic, ginger, and salt and mixed with water. The meat mixture is stuffed into the segment of gastrointestinal tract of yak, locally called *gyuma*, used as natural casings with 3–4 cm in diameter and 40–60 cm in length. One end of the casing is tied up with rope, and other end is sealed after stuffing, and the casing is boiled for 20–30 min. Cooked sausages are taken out and hung on bamboo strips above the kitchen oven for smoking and drying for 10–15 days or more to make *kargyong* (Figure 7.3). These are either dried or smoked, as per the producer's preference. The shelf life of *yak kargyong* is 2–3 months.

7.1.2.2 Culinary practices
Yak kargyong is consumed after boiling for 20–30 min, or it is sliced and fried in edible oil with onion, powdered or ground chillies, and salt. It is not sold in the market; it is generally prepared for home consumption (Figure 7.4).

7.1.2.3 Microorganisms
Bacteria: *Leuconostoc mesenteroides, Lactobacillus casei, Lb. plantarum, Lb. sake, Lb. divergens, Lb. carnis, Lb. sanfrancisco, Lb. curvatus, Enterococcus faecium, Bacillus subtilis,* and *B. mycoides*; yeasts: *Debaryomyces pseudopolymorphus* (Rai 2000).

Yak meat and ingredients

↓

◄— Add water

↓

Stuff into the intestine

↓

Tie both ends

↓

Boil for 3–5 min

↓

Air dry/smoke dry

↓

YAK KARGYONG

Figure 7.3 Indigenous method of *yak kargyong* preparation in Sikkim and Tibet.

Figure 7.4 Yak *kargyong* of Sikkim.

7.1.3   Faak kargyong

Faak kargyong is a traditional sausage prepared from pork meat in the Himalayas. It is commonly consumed by the Bhutia, Lepcha, Sherpa, Drukpa, and Tibetans.

7.1.3.1   Indigenous knowledge of preparation

Pork meat and fat are chopped, mixed with crushed garlic, ginger, soy sauce, and the required amount of salt and a small amount of water. The meat mixture is stuffed into the segment of gastrointestinal tract, locally called *gyuma*, used as natural casings with 3–4 cm in diameter and 40–60 cm in length. One end of the casing is tied up with rope, the other end is sealed after stuffing, and the casing is boiled for 20–30 min. Cooked sausages are stitched into bamboo strips above an earthen oven in the kitchen and smoked and dried for 10–15 days or more to make *kargyong* (Figure 7.5). Due to the use of natural casings, *faak kargyong* has a natural curved shape.

7.1.3.2   Culinary practices and economy

Faak kargyong is consumed as a curry after frying in edible oil, mixed with chilli, garlic, salt, and tomato. It is also sold in small hotels.

Faak kargyong is sold in the local market. It costs Rs. 170 per kilogram. It is also prepared during marriages and festivals and also for home consumption.

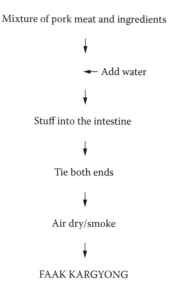

Mixture of pork meat and ingredients

↓

◄— Add water

↓

Stuff into the intestine

↓

Tie both ends

↓

Air dry/smoke

↓

FAAK KARGYONG

Figure 7.5 Indigenous method of *faak kargyong* preparation in Sikkim.

7.1.3.3   Microorganisms

Bacteria: *Lactobacillus brevis, Lb. carnis, Lb. plantarum, Leuconostoc mesenteroides, Enterococcus faecium, Bacillus subtilis, B. mycoides, B. licheniformis, Staphylococcus aureus,* and *Micrococcus* spp.; yeasts: *Debaryomyces polymorphus* and *Candida famata* (Rai 2008).

7.1.4   Lang satchu

Lang satchu is a threadlike or strand-like dried traditional beef meat product of the Himalayas. Tibetans, Drukpa, Bhutia, Lepcha, and Sherpa consume this product.

7.1.4.1   Indigenous knowledge of preparation

Red meat of beef is sliced into several strands of about 60–90 cm and is mixed thoroughly with turmeric powder, edible oil or butter, and salt. The meat strands are hung on bamboo strips or wooden sticks and are kept in the open air in a corridor of the house or are smoked above a kitchen oven for 10–15 days as per the preference of the consumers (Figure 7.6). *Lang satchu* can be kept at room temperature for several weeks. This is a natural preservation of perishable fresh raw meat in the absence of refrigeration or cold storage.

7.1.4.2   Culinary practices and economy

Lang satchu is made into a curry by washing and soaking in water briefly, squeezing out excess moisture, and frying with chopped garlic, ginger, chilli, and salt. Thick gravy is made and consumed with *thukpa* (noodles in soup) or boiled/baked potato. Deep-fried *satchu* is a popular side dish

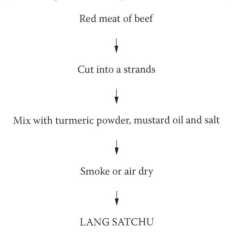

Figure 7.6 Indigenous method of *lang satchu* preparation in Sikkim.

in the local cuisine and is served with traditional alcoholic beverages/ drinks. It is also eaten directly.

Lang satchu is sold in the local restaurants and food stalls in Sikkim, the Darjeeling Hills, Bhutan, etc. A kilogram of *lang satch*u costs around Rs. 250–300. Some ethnic people are economically dependent upon this product.

7.1.4.3 Microorganisms

Bacteria: *Pediococcus pentosaceous, Lactobacillus casei, Lb. carnis, Entero-coccus faecium, Bacillus subtilis, B. mycoides, B. lentus, Staphylococcus aureus,* and *Micrococcus* spp.; yeasts: *Debaryomyces hansenii* and *Pichia anomala* (Rai 2008).

7.1.5 Yak satchu

Yak satchu is also a threadlike or strand-like dried traditional yak meat product of the Himalayas, mostly consumed by the Tibetans, Bhutia, Drukpa, and Lepcha.

7.1.5.1 Indigenous knowledge of preparation

Yak red meats are cut into several strands of about 60–90 cm and are mixed thoroughly with turmeric powder, edible oil or butter, and salt. The meat strands are hung on bamboo strips or wooden sticks and are kept in the open air in a corridor of the house or are smoked above a kitchen oven for 10–15 days as per the preference of the consumers (Figure 7.7). *Yak satchu* can be kept at room temperature for several weeks.

Red yak meat

↓

Cut into threads

↓

Mix with turmeric powder, mustard oil and salt

↓

Smoke or air dry

↓

YAK SATCHU

Figure 7.7 Indigenous method of *yak satchu* preparation in Sikkim.

Figure 7.8 Tibetan *satchu*.

7.1.5.2   Culinary practices and economy

Yak satchu is made into a curry by washing and soaking in water briefly, squeezing out excess water, and frying in yak butter with chopped garlic, ginger, chilli, and salt. Thick gravy is made and is eaten with noodles in soup or baked potatoes. Deep-fried *satchu* is a popular side dish of the ethnic people that is eaten with traditional alcoholic beverages in every house or on special occasions (Figure 7.8).

Yak satchu is not sold in the market. The ethnic people generally prepare the product for home consumption.

7.1.5.3   Microorganisms

Bacteria: *Pediococcus pentosaceous, Bacillus subtilis, B. mycoides, B. licheniformis, Streptococcus aureus,* and *Micrococcus* spp.; yeasts: *Debaryomyces polymorphus* (Rai 2008).

7.1.6   Suka ko masu

Suka ko masu is a dried or smoked meat product like *satchu*. It is dried or smoked meat prepared from buffalo meat or chevon (goat). It is commonly

Red meat (buffalo, chevon)

↓

Cut into threads

↓

Mix with turmeric powder, mustard oil and salt

↓

Smoke or air dry

↓

SUKA KO MASU

Figure 7.9 Indigenous method of *suka ko masu* preparation in Nepal.

consumed by the Nepali of the Darjeeling hills, Sikkim, and Nepal. The Newar community belonging to the Nepali calls it *sheakua*.

7.1.6.1   Indigenous knowledge of preparation

Suka ko masu is prepared by cutting the red meat of buffalo or chevon (goat meat) into strips up to 25–30 cm that are then mixed with turmeric powder, mustard oil, and salt. Mixed meat strips are hung on bamboo and kept above an earthen kitchen oven and smoked for 7–10 days. After complete drying, the smoked meat product is called *suka ko masu* (Figure 7.9), which can be stored at room temperature for several weeks.

7.1.6.2   Culinary practices and economy

Suka ko masu is washed and soaked in lukewarm water for 10 min, excess water is squeezed out, and the meat is fried in heated mustard oil with chopped onion, ginger, chilli powder, and salt. Coriander leaves are sprinkled over the curry, which is eaten with boiled rice. *Suka ko masu* is usually grilled in charcoal and is a popular side-dish in the region.

It is sold in the local markets at a cost of Rs. 350 per kilogram.

7.1.6.3   Microorganisms

Bacteria: *Lactobacillus carnis*, *Lb. plantarum*, *Enterococcus faecium*, *Bacillus subtilis*, *B. mycoides*, *B. thuringiensis*, *Staphylococcus aureus*, and *Micrococcus* spp.; yeasts: *Debaryomyces hansenii* and *Pichia burtonii* (Rai 2008).

7.1.7   Yak chilu

Yak chilu is a stored fat product prepared in North Sikkim, Tibet in China, and Bhutan. It is used in place of cooking oil during the scarcity of the same. It is mostly consumed by the Tibetans and Bhutia.

7.1.7.1   Indigenous knowledge of preparation

Fatty portions of freshly slaughtered meat (yak) are separated, kneaded by hand, and pressed into the cleaned and empty stomach of sheep (previously slaughtered) that is then stitched shut. This stuffed meat is pressed with heavy stones for about 5–10 h and is then kept hanging in a corridor of the house on a wooden plank for 10–15 days (Figure 7.10). *Chilu* can be used for a year or more. *Yak chilu* is not sold in the market; the people of North Sikkim usually prepare the product for home consumption.

Nowadays, this traditional method of preparation of *chilu* and *kheuri* is declining due to (a) the extinction of the sheep, locally known as *vyangloong* species, (b) the unavailability of sheep stomach, and (c) a government order banning the slaughtering of sheep.

7.1.7.2   Culinary practices

Yak chilu is used in place of edible oil for cooking by the Bhutia, Lepcha, Tibetans, Drukpa, etc.

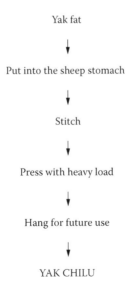

Yak fat

↓

Put into the sheep stomach

↓

Stitch

↓

Press with heavy load

↓

Hang for future use

↓

YAK CHILU

Figure 7.10 Indigenous method of *yak chilu* preparation in Sikkim.

7.1.7.3   Microorganisms
Unknown.

7.1.8   Lang chilu

Lang chilu is a stored fat product prepared from beef fat in North Sikkim, Tibet in China, and Bhutan. The Tibetans, Bhutia, and Lepcha use it in place of edible oil during the scarcity of such oil.

7.1.8.1   Indigenous knowledge of preparation
Fatty portions of freshly slaughtered meat (beef) are separated, kneaded by hand, and pressed into the cleaned and empty stomach of sheep (previously slaughtered) that is then stitched shut. This stuffed meat is pressed with heavy stones for about 5–10 h and is then kept hanging in a corridor of the house on a wooden plank for 10–15 days (Figure 7.11). *Lang chilu* can be used for a year or more.

Chilu production has declined due to the unavailability of sheep stomach. *Lang chilu* is used as a substitute for cooking oil.

7.1.8.2   Microorganisms
Unknown.

7.1.9   Luk chilu

Like other *chilu*, *luk chilu* is a stored sheep fat product prepared in North Sikkim, Tibet, and Bhutan. It is used in place of cooking oil during the scarcity of the edible oil.

Beef fat

↓

Put into the sheep stomach and stitch

↓

Press with heavy load and hang for future use

↓

LANG CHILU

Figure 7.11 Indigenous method of *lang chilu* preparation in Sikkim.

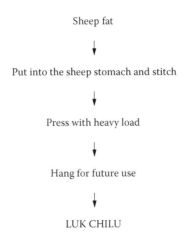

Figure 7.12 Indigenous method of *luk chilu* preparation in Sikkim.

7.1.9.1   Indigenous knowledge of preparation

The fat of freshly slaughtered mutton meat (sheep) is separated, kneaded by hand, and pressed into the cleaned and empty stomach of sheep (previously slaughtered) that is then stitched shut. The stuffed meat is pressed with heavy stones for about 5–10 h and is then kept hanging in a corridor of the house on a wooden plank for 10–15 days (Figure 7.12). *Luk chilu* can be preserved for a year or more. It is used as an edible oil.

7.1.9.2   Microorganisms

Unknown.

7.1.10   Yak kheuri

Yak kheuri is a typical indigenous yak meat product of the Himalayas and is consumed by Tibetans, Bhutia, Drukpa, and Lepcha. It is prepared during the winter season or whenever the raw materials are available.

7.1.10.1   Indigenous knowledge of preparation

During the preparation of *yak kheuri*, yak meat, intestines, and fat are chopped into pieces and mixed with the required amount of salt. The meat mixture is pressed into an empty stomach of sheep, locally called *khyabo* (previously cleaned sheep stomach). The stomach is stitched shut and kept for 1–2 months in the open air outside the kitchen for fermentation (Figure 7.13). Nowadays, the people of Sikkim have stopped preparing *yak kheuri* because of the unavailability of the sheep stomach due to a government ban on slaughtering high-altitude species of sheep.

Pieces of yak intestine (fatty), meat, fat

↓

◄— Add water

↓

Stuff ingredients into the sheep stomach and stitch

↓

Keep for 1 to 2 months in open air/ kitchen

↓

YAK KHEURI

Figure 7.13 Indigenous method of *yak kheuri* preparation in Sikkim.

7.1.10.2   *Culinary practices*

Yak kheuri is prepared by frying in yak butter, locally called *maa*, mixed with chopped ginger, onion, garlic, powdered or ground chillies, and salt and made into a thick curry. A simpler recipe involves boiling for 10–15 min with salt. *Yak kheuri* is consumed with main meals by the Tibetans, Bhutia, Drukpa, and Lepcha as a side dish or curry with baked potatoes.

   Yak kheuri is not sold in the local market. The ethnic people generally prepare the product for home consumption.

7.1.10.3   *Microorganisms*

Unknown.

7.1.11   *Lang kheuri*

Lang kheuri is an ethnic beef meat product of the Tibetans, Bhutia, and Lepcha in the Himalayas. It is prepared during the winter season or whenever the raw materials are available.

7.1.11.1   *Indigenous knowledge of preparation*

During the preparation of *lang kheuri*, beef meat, intestines, and fat are chopped into pieces and mixed with the required amount of salt. The meat mixture is pressed into an empty stomach of sheep, locally called *khyabo* (previously cleaned sheep stomach). The stomach is stitched shut and kept for 1–2 months in the open air outside the kitchen for fermentation (Figure 7.14). During the survey, it was observed that the people of North

Figure 7.14 Indigenous method of *lang kheuri* preparation in Sikkim.

Sikkim have stopped preparing *lang kheuri* because of the unavailability of the sheep stomach due to a government ban on slaughtering high-altitude species of sheep.

7.1.11.2 Culinary practices

Lang kheuri is prepared by frying in yak butter, locally called *maa*, mixed with chopped ginger, onion, garlic, powdered or ground chillies, and salt and made into a thick curry. A simpler recipe involves boiling for 10–15 min with salt. *Lang kheuri* is consumed with main meals by the Bhutia, Tibetans, and the Lepcha as a side dish or curry with baked potatoes.

  Lang kheuri is not sold in the local market. The ethnic people generally prepare the product for home consumption.

7.1.11.3 Microorganisms

Unknown.

7.1.12 Chartayshya

Chartayshya is a traditional dry or smoked chevon (goat) meat product consumed in the Kumaun hills of Uttarakhand.

7.1.12.1 Indigenous knowledge of preparation

Red goat meat (chevon) is cut into small pieces of 3–4 cm that are mixed with salt, sewed in a long thread, and hung on bamboo strips or wooden sticks in the open air in a corridor of the house for 15–20 days (Figure 7.15). It can be kept at room temperature for several weeks for future consumption

Raw chevon meat

↓

Cut into small pieces; marinate with salt

↓

Smoke or dry for 15–20 d

↓

CHARTAYSHYA

Figure 7.15 Indigenous method of *chartayshya* preparation in Kumaun.

Figure 7.16 *Chartayshya* of Kumaun hills.

(Figure 7.16). In western Nepal, a similar product called *sukha sikhar* is prepared from chevon.

7.1.12.2 Culinary practices and economy
Curry is made by frying in edible oil with tomato, ginger, garlic, onion, and salt. It is not sold in the local market.

7.1.12.3   Microorganisms

Bacteria: *Lactobacillus divergens, Enterococcus faecium, Pediococcus pentosaceous, Bacillus subtilis, B. mycoides, B. thuringiensis, Staphylococcus aureus,* and *Micrococcus* spp.; yeasts: *Debaryomyces hansenii* and *Candida famata* (Rai 2008).

7.1.13   Jamma/geema

Jamma or *geema* is a traditional fermented sausage (Figure 7.17) of the Kumaun Himalayas prepared from chevon meat. These products are also consumed by the Bhutia of Dharchula and Munsiari of Pithorgarh district of Uttarakhand.

7.1.13.1   Indigenous knowledge of preparation

Red goat meat is chopped into fine pieces, and ground finger millet (*Eleusine coracana*), wild pepper (locally called *timbur* [*Zanthoxylum* sp.]), chilli powder, and salt are added and mixed. A small amount of fresh animal blood is also added. The meat mixture is made semi-liquid by adding water, and the slurry is stuffed through a funnel into the small intestine of goat of about 2–3 cm in diameter and 100–120 cm in length. Both ends

Figure 7.17 *Jamma* of Kumaun hills.

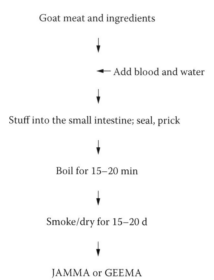

Goat meat and ingredients

↓

◄— Add blood and water

↓

Stuff into the small intestine; seal, prick

↓

Boil for 15–20 min

↓

Smoke/dry for 15–20 d

↓

JAMMA or GEEMA

Figure 7.18 Indigenous method of *jamma* preparation in Kumaun.

of the small intestine are tied off, and the casing is boiled for 15–20 min. During the boil, the stuffed intestines are pricked randomly to prevent bursting. After boiling for 15–20 min, the stuffed intestines are smoked above a kitchen oven for 15–20 days (Figure 7.18) or they can be eaten as such. This food is similar to *kargyong* of Sikkim and Bhutan.

7.1.13.2   Culinary practices
Jamma is consumed as a curry by mixing with onion, garlic, ginger, tomato, and salt. It is also deep fried and is eaten with local alcoholic beverages. Sometimes, *jamma* may be eaten as cooked sausage. It is not sold in the local market.

7.1.13.3   Microorganisms
Bacteria: *Leuconostoc mesenteroides, Lactobacillus sanfrancisco, Lb. divergens, Pediococcus pentosaceous, Enterococcus cecorum, E. faecium, Bacillus subtilis, B. sphaericus, Staphylococcus aureus,* and *Micrococcus* spp.; yeasts: *Debaryomyces hansenii* and *Candida albicans* (Rai 2008).

7.1.14   Arjia

Arjia is an ethnic sausage prepared from chevon (goat) meat. It is also an important food of the Kumaun.

Goat meat, goat lungs, chilli, garlic

↓

◄— Add blood and water

↓

Stuff into the large intestine; boil for 15–20 min

↓

Dry/smoke for 15–20 d

↓

ARJIA

Figure 7.19 Indigenous method of *arjia* preparation in Kumaun.

7.1.14.1   Indigenous knowledge of preparation

The preparation method of *arjia* is similar to that of *jamma*. In this case, however, a mixture of chopped lungs of goat, salt, chilli powder, *timbur* (*Zanthoxylum* sp.), and fresh animal blood are stuffed into the large intestine of goat, instead of the small intestine, and boiled for 15–20 min. Pricking of stuffed large intestine is necessary to prevent bursting while boiling. The boiled casing is dried or smoked for 15–20 days above a kitchen oven (Figure 7.19).

7.1.14.2   Culinary practices

Arjia is consumed as a curry or as a deep-fried sausage along with main meal. It is not sold in the local market.

7.1.14.3   Microorganisms

Bacteria: *Enterococcus faecium, Pediococcus pentosaceous, Bacillus subtilis, B. mycoides, B. thuringiensis, Staphylococcus aureus,* and *Micrococcus* spp.; yeasts: *Debaryomyces hansenii* and *Cryptococcus humicola* (Rai 2008).

7.2   Microbiology

Viable counts of LAB in the Himalayan meat products are above 10^8 cfu/g, followed by Micrococcaceae and yeasts, with populations not exceeding 10^7 cfu/g (Table 7.1). Micrococcaceae, LAB, and yeasts are the typical microflora in meat products, with the involvement of Micrococcaceae being the

Table 7.1 Microbiological Populations of Himalayan Meat Products

| | Log cfu/g sample | | | | | |
| | Bacteria | | | | | |
Product	LAB[a]	Bacilli	Micrococcaceae[b]	Yeast	Molds	AMC[c]
Lang kargyong	6.5 ± 0.4	1.8 ± 0.4	5.6 ± 0.4	5.6 ± 3.4	1.2 ± 0.2	8.3 ± 0.3
Yak kargyong	6.4 ± 0.5	1.7 ± 0.4	5.6 ± 0.3	4.9 ± 0.1	2.1 ± 0.1	8.0 ± 0.5
Faak kargyong	7.1 ± 0.2	1.5 ± 0.3	6.0 ± 0.3	5.3 ± 0.3	0	8.2 ± 0.2
Lang satchu	7.1 ± 0.1	1.7 ± 0.5	6.5 ± 0.3	5.1 ± 0.2	0	8.1 ± 0.1
Yak satchu	6.4 ± 0.3	1.7 ± 0.4	5.8 ± 0.6	4.6 ± 0.4	0	7.8 ± 0.2
Suka ko masu	6.0 ± 0.1	2.5 ± 0.1	4.9 ± 0.1	4.4 ± 0.1	0	6.2 ± 0.1
Chartayshya	7.0 ± 0.2	2.8 ± 0.1	6.6 ± 0.1	4.9 ± 0.1	1.6 ± 0.1	7.8 ± 0.1
Jamma	7.4 ± 0.1	3.5 ± 0.1	5.5 ± 0.1	5.7 ± 0.1	3.5 ± 0.1	7.8 ± 0.1
Arjia	7.8 ± 0.1	3.8 ± 0.1	6.4 ± 0.1	4.7 ± 0.1	0	9.0 ± 0.1

Note: Data represent the means (± SD).
[a] LAB, lactic acid bacteria.
[b] Micrococcaceae includes species of *Micrococcus* and *Staphylococcus*.
[c] AMC, aerobic mesophilic count.

most important (Vilar et al. 2000). Diverse species of LAB have been isolated from *lang kargyong, yak kargyong, faak kargyong, lang satchu, yak satchu, suka ko masu, chartayshya, jamma,* and *arjia* food samples of the Himalayas (Rai 2008).

Classification of LAB into different genera is largely based on morphology, gas production from glucose (Kandler 1983), mode of glucose fermentation, and growth at different temperatures (Mundt 1986; Dykes et al. 1994). Species of LAB are identified as *Lb. sake, Lb. curvatus, Lb. divergens, Lb. carnis, Lb. sanfrancisco, E. faecium, E. cecorum, Leuconostoc mesenteroides, Lb. plantarum, Lb. casei, Lb. brevis,* and *P. pentosaceous* (Rai 2008). The presence of *Lb. divergens, Lb. carnis, Lb. sake,* and *Lb. curvatus* has been reported in vacuum-packaged meat (Schillinger and Lücke 1989). *Leuconostoc mesenteroides, P. pentosaceous, Lb. brevis,* and *E. faecium* have also been isolated from other meat products (Samelis et al. 1994).

Lactobacillus sake isolated from *lang kargyong* and *yak kargyong* does not ferment maltose and lactose; this has also been observed by other workers (Schillinger and Lücke 1987; Hastings and Holzapfel 1987a, 1987b; Grant and Patterson 1991). Enterococci play a beneficial role in production of many fermented foods (Bouton et al. 1998; Cintas et al. 2000). *Enterococcus faecium* appears to pose a low risk for use in foods, because these strains generally harbor fewer recognized virulence determinants than *E. faecalis* (Franz et al. 2003). Santos et al. (1998) isolated *Pediococcus* sp. from a Spanish dry-fermented sausage, chorizo. Several species and strains of pediococci are used as starter cultures in fermentation of meat (Smith and Palumbo 1983) and sausage products (Tagg et al. 1976).

Among Micrococcaceae, about 91% of the isolates are identified as *Staphylococcus* spp. and only 9% are *Micrococcus* spp. in 68 samples of ethnic meat products of the Himalayas showing dominance of *Staphylococcus aureus* and other spp. over *Micrococcus* spp. (Rai 2008). There is a predominance of staphylococci over other Micrococcaceae in fermented meats (Samelis et al. 1998), raw-cured meat products (Rodriguez et al. 1994), and fermented sausages (Papamanoli et al. 2002). Micrococcaceae species are used to enrich fermentative microorganisms during aging of the products in order to enhance the color stability of the cured meat and prevent rancidity (Papamanoli et al. 2002). The activities of the Micrococcaceae group reduce spoilage, decrease processing time, and contribute to flavor development (Montel et al. 1998). Considering these reports, native Micrococcaceae in the Himalayan meat products is essential in the final product development.

Bacilli are also detected at the level of less than 10^4 cfu/g for the Himalayan meat products. The results for the technological properties of *Bacillus* strains isolated from southern Italian sausage made without a selected starter suggest that *Bacillus* strains, always present in meat curing, could play a role in the development of texture and organoleptic characteristics of the sausages (Baruzzi et al. 2006).

Based on the detailed characterization and identification profiles, the following yeasts are isolated from the Himalayan meat products and identified as *Debaryomyces hansenii, D. polymorphus, D. pseudopolymorphus, Pichia burtonii, P. anomala, Candida famata, C. albicans,* and *C. humicola.* Species of *Debaryomyces* are the dominant yeasts in the Himalayan meat products (Rai 2008). Although bacteria are considered to have the dominant role in meat fermentation, the contribution of yeasts nevertheless is significant (Romano et al. 2006). A diversity of yeast species has been isolated from fermented sausages and cured hams produced in different countries with little exception (Tamang and Fleet 2008). Simoncini et al. (2007) have reported the occurrence of *D. hansenii* and *Candida famata* from Italian dry cured ham. The presence of *D. hansenii* in the Himalayan meat products may contribute in color and flavor development during the traditional processing. Curing color and flavor could be improved by addition of selected *Debaryomyces* strains to the sausage mixture (Tamang and Fleet 2009).

7.2.1 Prevalence of LAB

The most dominant LAB in *lang kargyong* are lactobacilli (71.7%) followed by *Enterococcus* spp. (15.2%) and *Leuconostoc* spp. (13.1%) (Rai 2008).

Similarly, in the case of *yak kargyong,* lactobacilli are the dominant micro-flora (81.1%), followed by *Leuconostoc* spp. (11.3%) and *Enterococcus* spp. (7.6%). *Leuconostoc* spp. are the dominant LAB (52.6%) in *faak kargyong,* followed by *Enterococcus* spp. (35.9%) and lactobacilli (11.5%). In *lang satchu,* the prevalence of *Enterococcus* spp. is 74.2%, followed by *Pediococcus* sp. (23.7%) and lactobacilli (2.1%). *Pediococcus* sp. is the most dominant lactic microflora, representing 100% in *yak satchu,* whereas in *suka ko masu,* the dominant LAB are *Enterococcus* spp. (72.7%), followed by lactobacilli (27.3%). The most dominant LAB recovered from *chartayshya* is *Pediococcus* sp. (56.9%), followed by *Enterococcus* spp. (29.3%) and lactobacilli (13.8%). About 47.8% of LAB is represented by *Leuconostoc* spp., 21.7% by *Pediococcus* sp., 18.9% by lactobacilli, and 11.6% by *Enterococcus* spp. in *jamma.* The dominant LAB in *arjia* are *Enterococcus* spp., representing 54.2% of the total microorganisms, followed by 45.8% of *Pediococcus* sp.

7.2.2   Occurrence of pathogenic bacteria

Meat is highly or extremely susceptible to microbial spoilage (Farnworth 2003). Certain faults in the manufacture and storage of the meat products may lead to outbreaks of food-borne pathogens (Lücke 2003). Fermented meat products usually do not undergo a physical treatment to eliminate pathogenic microorganisms; the meat has to be high quality with regard to hygienic and microbial counts, and the control of pathogens is achieved by appropriate fermentation technology. Interestingly, no food-borne pathogens—*Bacillus cereus, Salmonella, Shigella, Listeria monocytogenes*—have been isolated from the Himalayan sausages and meat products (Rai 2008). This shows that the traditionally processed ethnic meat products of the Himalayas are safe for eating. Fermented meat and sausages are considered safe for consumption due to low pH and water activity, which inhibit the growth of pathogenic bacteria (Ferreira et al. 2006; Aymerich 2008).

7.2.3   Technological or functional properties

LAB strains isolated from the Himalayan meat products show weak lipolytic activity (Rai 2008). In *nham,* a fermented meat product of Thailand, weak lipolytic activity of LAB strains has been observed (Montel et al. 1998). The absence of proteinases (trypsin and chymotrypsin) and the presence of strong peptidase (leucine-, valine-, and cystine-arylamidase) activities is exhibited by the LAB strains isolated from meat products of the Himalayas. Proteolytic events that take place during the processing of dry sausages result in an increase in small peptides and free amino acids, thereby contributing to the overall flavor in cured meat products such as dry sausages (Demeyer et al. 2000). Nearly all of the strains of LAB isolated from meat products (with a few exceptions) show strong antimicrobial activity against a number of potentially pathogenic Gram-positive and Gram-negative bacteria (Rai 2008). However, LAB strains of the Himalayan meat products could not produce bacteriocin. Production of bacteriocin depends on a number of intrinsic and extrinsic factors including redox potential, water activity, pH, and temperature (Delgado et al. 2005). It could be speculated that the antimicrobial activities shown by the strains isolated from the Himalayan meat products may be due to the other antimicrobials, such as organic acids, hydrogen peroxide, etc. Ammor et al. (2006, 2007) explain that *Lb. sake* isolate displayed an additional inhibitory effect by hydrogen peroxide against *Listeria innocua*.

As in other fermented foods, biogenic amines are also present in dry sausages (ten Brink et al. 1990; Halász et al. 1994). Samples with moderate, high, or very high levels of biogenic amines can be considered as products of lower quality, and their consumption could be unhealthy for sensitive individuals (Latorre-Moratalla et al. 2007). None of the microorganisms from the Himalayan meat products are found to produce biogenic amines (Rai 2008). Nonproduction of biogenic amine by strains of LAB in ethnic sausages and other meat products of the Himalayas is a good property to be selected as a starter culture.

Microorganisms isolated from the Himalayan meat products have also been tested for their adherence ability to ascertain their probiotic characteristics. Only two strains, *Lb. brevis* and *Lb. plantarum*, isolated from *faak kargyong* and *yak kargyong*, respectively, had more than 70% hydrophobicity, indicating their probiotic character (Rai 2008).

It is interesting to observe that native microorganisms from the Himalayan sausages and other meat products have several technological properties contributing functionally to product quality. Selection of starter culture for production of sausages and other meat products from the native LAB may be suggested for better quality in term of sensory, food safety, and functional aspects.

7.3   Nutritive value

The nutritional value of Himalayan meat products is shown in Table 7.2. The pH of all products is slightly acidic in nature due to the predominance of the LAB flora. As a result of dehydration *yak kargyong, faak kargyong, lang satchu, yak satchu, suka ko masu*, and *chartayshya* have a low moisture content (Rai 2008). Due to the low moisture content and slightly acidic nature of some of the meat products, such as *kargyong, satchu, suka ko masu, chartayshya*, etc., can be kept for a longer period and are safe for consumption. Due to their high calorie value, the Himalayan ethnic meat products are a good source of animal protein and other nutritional intake for the people living in high mountainous regions.

Table 7.2 Nutritional Composition of Himalayan Meat Products

Parameter[a]	Ethnic meat products								
	Lang kargyong	Yak kargyong	Faak kargyong	Lang satchu	Yak satchu	Suka ko masu	Chartayshya	Jamma	Arjia
Moisture (%)	59.8 ± 0.8	21.9 ± 0.7	41.0 ± 1.0	22.8 ± 0.7	23.7 ± 1.5	23.2 ± 0.3	17.4 ± 0.2	65.1 ± 0.6	60.2 ± 0.2
pH	6.7 ± 0.3	6.9 ± 0.2	6.5 ± 0.3	6.1 ± 0.2	5.7 ± 0.2	5.3 ± 0.3	6.5 ± 0.1	5.5 ± 0.2	6.3 ± 0.1
Acidity (% as lactic acid)	0.3 ± 0.1	0.3 ± 0.1	0.6 ± 0.2	2.6 ± 0.2	2.5 ± 0.2	1.1 ± 0.2	2.1 ± 0.1	1.5 ± 0.1	1.1 ± 0.1
Ash (% DM)	3.8 ± 0.6	2.8 ± 0.4	2.8 ± 0.3	5.4 ± 0.6	7.3 ± 0.4	1.8 ± 0.4	7.8 ± 1.2	5.2 ± 0.5	3.5 ± 0.4
Fat (% DM)	10.3 ± 1.1	49.1 ± 0.7	27.1 ± 0.8	5.9 ± 0.6	4.7 ± 0.5	2.0 ± 0.1	17.0 ± 0.2	4.2 ± 0.5	5.5 ± 0.4
Protein (% DM)	8.4 ± 1.2	16.0 ± 0.8	11.5 ± 0.8	57.7 ± 1.2	51.0 ± 1.6	44.8 ± 0.8	36.6 ± 3.0	7.8 ± 1.0	6.4 ± 0.9
Carbohydrate (% DM)	77.5 ± 2.7	32.0 ± 1.7	58.6 ± 2.1	31.0 ± 2.3	37.0 ± 2.4	51.4 ± 1.5	38.6 ± 4.3	82.8 ± 2.0	84.6 ± 1.7
Food value (kcal/100 g DM)	436.2 ± 3.2	634.5 ± 2.3	501.4 ± 4.8	407.7 ± 2.0	405.8 ± 0.9	403.1 ± 0.5	454.0 ± 5.9	400.0 ± 0.7	413.5 ± 1.2

Note: Data represent the mean values.

[a] DM, dry matter.

7.4   Conclusion

Drying, smoking, and fermentation are traditional techniques for processing highly perishable raw meats in the Himalayas. These techniques are remarkable for their biopreservation efficiency. Preparation of sausages is different from that of the European method in regard to the use of natural casing, meat mixture, and particularly the sausages of yak product, which are unique to the Himalayas. These meat products are preserved for several months without refrigeration and can be consumed at any time. The dominant microorganisms in the Himalayan sausages and meat products are lactic acid bacteria, followed by yeasts, Micrococcaceae, and bacilli. Some of the strains of LAB have functional properties that may be selected as a starter culture for optimized production of sausages typical of the Himalayas.

chapter eight

Ethnic starters and alcoholic beverages

The malting process for alcohol production in the Himalayas is rarely or never used in traditional alcohol fermentation processes. Wine making is also not a tradition in the Himalayas, since fruits are eaten directly without extracting them into juice or fermenting them into wines. Instead, dry-mixed starters containing a consortium of microorganisms are traditionally used in the fermentation of alcoholic beverages in the Himalayas.

In Asia, three major types of inocula as starters are commercially produced to convert starchy materials to sugars and subsequently to alcohol and organic acids (Hesseltine 1985b; Tamang and Fleet 2009). In the first type, pure cultures of *Aspergillus oryzae* and *Aspergillus sojae* are used in the form of a starter (called *koji* in Japan). At the same time, they produce amylases that convert starch to fermentable sugars, which are then used for the second-stage yeast fermentation to make miso and shoyu (fermented soybean products), while proteases are formed to break down the soybean protein. In the second type, whole-wheat flour with its associated flora is moistened and made into large compact cakes, which are incubated to select certain desirable organisms. The cakes, after a period of incubation, are used to inoculate large masses of starchy material, which are then fermented to produce alcohol. The cakes contain yeasts and filamentous molds. This inoculum is used in the so-called *kao-liang* process for making alcohol. In the third type, the starter is a mixed culture of yeast, filamentous molds, and bacteria. This starter is in the form of flattened or round balls of various sizes, compact in texture, and dry. The starter is inoculated with some previous starter. This microflora mixture is allowed to develop for a short time, then dried and used to make either alcohol or fermented foods from starchy materials. The use of mixed starters in the form of dry ball-like cakes containing amylolytic and alcohol-producing yeasts, starch-degrading molds, and lactic acid bacteria is common in many Asian countries (Nout and Aidoo 2002; Dung et al. 2006a; Tamang and Fleet 2009).

Alcoholic drinks have been widely consumed since pre-Vedic times in India (Prakash 1987). Ethnic alcoholic beverages are exclusively prepared from locally grown cereal grains such as finger millet, rice, maize, barley, wheat, and other starchy materials using traditionally prepared dry-mixed inocula or starter in the form of a flattened cake or ball in the Himalayas. Ethnic alcoholic beverages have strong ritualistic importance among the ethnic people in the Himalayas, where social activities require the provision and consumption of appreciable quantities of alcohol (Tamang et al. 1996). Traditional brewing of alcohol is a home-based industry mostly practiced by rural mountain women in the Himalayas using their native skills of alcohol fermentation. Hutchinson and Ram-Ayyar (1925) have described the Indian rice starter culture called *bakhar* and its alcoholic products.

This chapter describes various types of ethnic mixed starters used for alcohol fermentation (*marcha, manapu, mana, hamei, khekhrii*), ethnic alcoholic beverages (*kodo ko jaanr/chyang, bhaati jaanr, poko, makai ko jaanr, gahoon ko jaanr, atingba, zutho, sura*), and traditional distilled liquor or alcoholic drinks (*raksi*) of the Himalayas.

8.1   Traditional starter culture

8.1.1   Marcha

Marcha or *murcha* (collectively spelled *marcha*) is a ball-like starter, used to ferment starchy materials into fermented beverages in Nepal, Bhutan, Darjeeling hills, and Sikkim (Figure 8.1). *Marcha* or *murcha* is not a food item; rather, it is a mixed-dough inoculum used as a starter culture for

Figure 8.1 *Marcha* or *phab* of the Himalayas.

the preparation of various indigenous alcoholic beverages. *Marcha* is a dry, round to flat, creamy white to dusty white, solid ball-like starter ranging from 1.9 to 11.8 cm in diameter, with the weight ranging from 2.3 to 21.2 g. *Marcha* is the Nepali word. *Marcha* has various vernacular names: *phab* (Tibetans and Bhutia), *khesung* (Limboo), *bharama* (Tamang), *bopkha* or *khabed* (Rai), *buth/thanbum* (Lepcha), *poo* (Drukpa), and *manapu* and *mana* (Newar).

8.1.1.1 Indigenous knowledge of preparation

At the first stage of *marcha* preparation, glutinous rice (*Oryza sativa*) is soaked in water for 8–10 h at ambient temperature. Unheated soaked rice is crushed in a foot-driven, heavy wooden mortar and pestle. In 1 kg of ground rice, ingredients added include roots of *guliyo jara* or *chitu* (*Plumbago zeylanica*), 2.5 g; leaves of *bheemsen paate* (*Buddleja asiatica*), 1.2 g; flowers of *sengreknna* (*Vernonia cinerea*), 1.2 g; ginger, 5.0 g; red dry chilli, 1.2 g; and previously prepared *marcha* as mother culture, 10.0 g. The mixture is then made into a paste by adding water and kneaded into flat cakes of varying sizes and shapes. This is then placed individually on the ceiling–floor, above the kitchen, made up of bamboo strips inlaid with fresh fronds of ferns, locally called *pire uneu* (*Glaphylopteriolopsis erubescens*), and covered with dry ferns and jute bags (Figure 8.2). These are left to ferment for 1–3 days, depending upon the temperature. Completion of fermentation is indicated by a distinct alcoholic and ester aroma and puffy/swollen appearance of the *marcha*. Finally, cakes of *marcha* are sun dried for two to

Figure 8.2 A Limboo woman making *marcha* in Nepal.

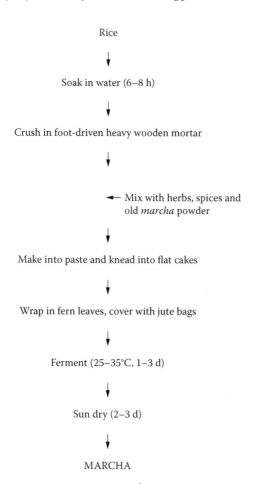

Figure 8.3. Indigenous method of *marcha* preparation in East Sikkim.

three days (Figure 8.3). *Marcha* is stored at room temperature and in a dry place for more than a year.

In north Sikkim, root barks and flowers of wild herbs, locally called *marcha jar* (*Polygala arillata*), are mixed and ground with water during *marcha* preparation (Thapa 2001). In eastern Nepal, the *marcha* producer uses more than 42 plants (2 ferns, 5 monocots, and 35 dicots) and their roots and leaves for making *marcha*. The more widely used plants (called *marcha* plants) are *Vernonia cinerea, Clematis grewiaeflora, Polygala arillata, Piper chaba, P. longum, Plumbago zeylanica, Buddleja asiatica, Christella appendiculata, Polygala* sp., *Elephantopus scaber, Inula* sp., and *Scoparia dulcis* (KC et al. 2001).

Phab (similar to *marcha*) is prepared by the Khampa community of Tibet (Bhatia et al. 1977). The important places of *phab* production are Nubra in

Ladakh and Manali in Himachal Pradesh, and its formulation and art of preparation is a closely guarded secret with the Tibetan-Khampa (Bhatia et al. 1977).

8.1.1.2   Interpretation of ethno-microbiology

The native skill of alcohol production by starter-culture technique, coupled with the use of traditional distillation apparatus, is well recognized. The consortium of microorganisms is preserved in a rice or wheat base, a source of starch (a "medium," in modern terms), and glucose-rich wild herbs are used to supplement the carbon source for growing microorganisms. This consortium of microorganisms is called a "starter culture" in the field of modern food science.

Unlike mixed culture starters of the other Asian countries, *marcha* is usually prepared by wrapping kneaded dough cakes in fern fronds, with the fertile side touching them. This may be due to an abundance of ferns, locally called *pire uneu* (*Glaphylopteriolopsis erubescens* [Wall ex Hook.] Ching), in the Himalayas. Probably, germination of spores in sori helps to maintain the warmth of the fermenting mass in cold climates. Preparation of *marcha* is similar to that of other starter cultures of Asia. *Marcha* makers believe that addition of wild herbs give more sweetness to the product, and they also believe that addition of chillies and ginger during *marcha* preparation is to get rid of devils that may spoil the product. This is actually to prevent growth of undesirable microorganisms that may inhibit the growth of native functional microorganisms in *marcha*. Studies of Soedarsono (1972) in *ragi*, an Indonesian rice-based starter culture, reveal that certain spices inhibit many undesirable microorganisms at the time of fermentation. Hesseltine (1983) has speculated that the spices, which are known to be inhibitory to many bacteria and molds, are the agents that select the right population of microorganisms for fermentation. *Marcha*-making technology reflects the traditional method of subculturing desirable inocula from previous batch to new culture using rice as base substrates. This technique preserves the microbial diversity essential for beverage production. *Marcha* retains its potency in situ for over a year or more.

8.1.1.3   Socioeconomy

Annual production of *marcha* in rural Sikkim is 101 kg per house (Tamang et al. 2007a). *Marcha* is produced at home for commercial use in a few *marcha*-making villages in eastern Nepal, Darjeeling hills, Sikkim, and Bhutan. Men help women in collecting wild herbs and pounding them during *marcha* preparation. *Marcha* is exclusively prepared by the rural women belonging to the Limboo, Rai, and Lepcha in Darjeeling hills and Sikkim, and in Nepal by the Limboo, Rai, Tamang, Gurung, Majhi, Kumhal, Dhimal, etc. This art of technology is protected as a hereditary trade and passes from mother to daughter. The *marcha*-making villages have linkages to nearby markets where *marcha* makers can sell the products once or twice a week to supplement household income.

Some villages in the Eastern Himalayas in Darjeeling hills, Sikkim, and East Nepal are central resources for the microbes or genes involved in mountain microbiological systems for *marcha* making. During our survey, these villages were identified as the possible microbial resources necessary for ethnic alcohol production. The Himalayan women have been subculturing and maintaining a consortium of functional microorganisms for alcohol production in the form of a dry, flattened-ball, cake-like starter called *marcha* or *phab* for more than 2000 years. Some of the important *marcha*-making villages in the Eastern Himalayas are: Nor Busty (Darjeeling: dominant *marcha* makers are Rai and Limboo women); Kashyong (Kalimpong: dominant *marcha* makers are Rai and Limboo women); Mangzing (Kalimpong: dominant *marcha* makers are Limboo, Rai, and Lepcha women); Jhosing (North Sikkim: dominant *marcha* makers are Limboo women); Tibuk (North Sikkim: Limboo); Chhejo (West Sikkim: Limboo); Lingchom (West Sikkim: Limboo); Salghari (South Sikkim: Limboo); Barnyak (South Sikkim: Limboo, Rai); Aho (East Sikkim: Limboo); Kopchey (East Sikkim: Limboo); Change (Taplejung-Nepal: Limboo); Ahale (Bhojpur-Nepal: Tamang); Bokhim (Bhojpur-Nepal: Gurung); Rajarani (Dhankuta-Nepal: Limboo); Nundhaki (Sankhuwasabha-Nepal: Limboo); Terhathum (Terhathum-Nepal: Limboo). These *marcha*-making villages should be preserved, and the rural women involved should be trained more on the microbiology of their processes while acknowledging their knowledge of subculturing and preserving the necessary functional microorganisms.

The cost of *marcha* depends on its shape and size; small pieces of *marcha* (≈2.0 cm) cost about 50 paisa each, while large pieces (≈12 cm) sell for Rs. 5 (Figure 8.4). Earnings from the sale of *marcha* are an important source of income for the producers, who earn a profit of about 60%–70%. The production and sale of *marcha* is constantly increasing as an unorganized industry sector. However, none of the governments in the Himalayan regions have officially declared it as a trade. *Marcha* producers are definitely contributing to the regional economy, which should be recognized by the federal and local governments.

8.1.1.4   Similar traditional starters

Marcha is similar to the traditional mixed starter cultures used in other countries in Southeast Asia. These starter cultures are used to prepare alcoholic beverages from starchy substrates such as *chu* or *chou* of China (Wei and Jong 1983; Yokotsuka 1991), *ragi* of Indonesia (Saono et al. 1974), *nuruk* of Korea (Kim 1968), *bubod* of the Philippines (Tanimura et al.

Figure 8.4 Varieties of *marcha* are sold in local markets in the Eastern Himalayas.

1978; Kozaki and Uchimura; 1990), *loogpang* of Thailand (Vachanavinich et al. 1994), and *men* of Vietnam (Dung et al. 2006a).

Several other traditional mixed starter cultures similar to *marcha* are used for alcohol production in different parts of the Himalayas. These are *hamei* of Manipur; *manapu* and *mana* of Nepal; *phab* of Tibet, Ladakh; *chang-poo* of Bhutan; *pham* and *ipoh* of Arunachal Pradesh; *bakhar* of Himachal Pradesh; *emao* of Assam; *thiat* of Meghalaya; *pham* of Arunachal Pradesh; and *balan* of Uttarakhand.

8.1.1.5   Microorganisms

Filamentous molds: *Mucor circinelloides, M. hiemalis, Rhizopus chinensis,* and *R. stolonifer* var. *lyococcus;* yeasts: *Saccharomycopsis fibuligera, Saccharomycopsis capsularis, Pichia anomala, P. burtonii, Saccharomyces cerevisiae, S. bayanus,* and *Candida glabrata;* LAB: *Pediococcus pentosaceus, Lactobacillus bifermentans,* and *Lb. brevis* (Tamang 1992; Tamang and Sarkar 1995; Thapa 2001; Tsuyoshi et al. 2005; Tamang et al. 2007a).

8.1.2   Manapu

Manapu is a mixed starter, the same as *marcha,* that is prepared from rice flour and millet grains in Nepal (Karki 1994; Shrestha et al. 2002).

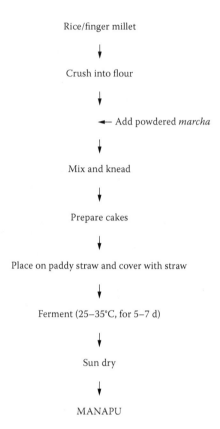

Figure 8.5 Indigenous method of *manapu* preparation in Nepal.

8.1.2.1 Indigenous knowledge of preparation

Rice or millet is milled to obtain flour and then mixed with 20% old *marcha* (as inoculum), 5% *manawasha* (white flower of a wild plant), and 5% black pepper. It is then kneaded to prepare a cake and placed on straw, which is then covered by more straw and fermented at 30°C–33°C for 5–7 days. After completion of fermentation, fresh cakes are sun dried to get the final *manapu* (Figure 8.5).

8.1.2.2 Microorganisms

Yeasts: *Saccharomyces cerevisiae* and *Candida versatilis*; mold: *Rhizopus* spp.; LAB: *Pediococcus pentosaceus* (Shrestha et al. 2002).

8.1.3 Mana

Mana is a granular-type starter prepared from wheat flakes in Nepal (Shrestha et al. 2002).

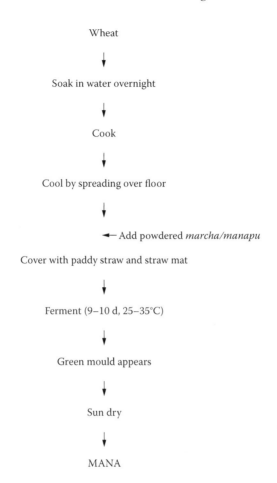

Wheat

↓

Soak in water overnight

↓

Cook

↓

Cool by spreading over floor

↓

◄— Add powdered *marcha/manapu*

Cover with paddy straw and straw mat

↓

Ferment (9–10 d, 25–35°C)

↓

Green mould appears

↓

Sun dry

↓

MANA

Figure 8.6 Indigenous method of *mana* preparation in Nepal.

8.1.3.1   *Indigenous knowledge of preparation*

Wheat grains are soaked in water overnight and steamed for 30 min. The broth is then transferred to a bamboo basket and drained, and the softened wheat grains are ground using a traditional mortar and pestle until they become lumpy. The floor is cleaned, straw is spread on ground, and the wheat slurry is placed over it, covered with paddy straw or a straw mat, and fermented for 6–7 days. After 7 days, a green mold appears on the wheat grains. The moldy wheat is dried in the sun to get *mana*, which is stored for future use (Figure 8.6).

8.1.3.2   *Microorganisms*

Molds: *Aspergillus oryzae* (Nikkuni et al. 1996) and *Rhizopus* spp. (Shrestha et al. 2002).

Figure 8.7 *Hamei* of Manipur.

8.1.4   *Hamei*

Hamei is a dry, round-to-flattened, solid ball-like, mixed-dough inoculum used as a starter culture for preparation of various indigenous alcoholic beverages in Manipur (Figure 8.7).

8.1.4.1   *Indigenous knowledge of preparation*

During traditional production of *hamei*, local varieties of raw rice, without soaking or after soaking for 30 min and then dried, are crushed and mixed with powdered bark of *yangli* (*Albizia myriophylla* Benth.) and a pinch of previously prepared powdered *hamei*. The dough is pressed into flat cakes by palms, which determine the shape, sizes, and forms of *hamei* as per the choice of the *hamei* makers. The pressed cakes are kept over paddy husk or paddy straw in a bamboo basket, covered by sackcloths for 2–3 days at room temperature (20°C–30°C), and then sun dried for 2–3 days (Figure 8.8). *Hamei* is used to prepare a rice-based beverage, locally called *atingba*, and a distilled clear liquor called *yu* in Manipur. The good quality of *hamei* increases the alcohol content in *yu* (Singh and Singh 2006).

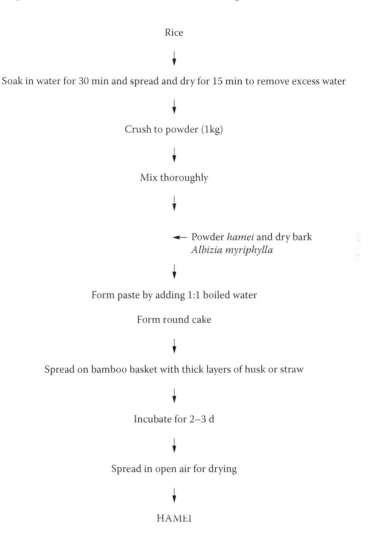

Rice

↓

Soak in water for 30 min and spread and dry for 15 min to remove excess water

↓

Crush to powder (1kg)

↓

Mix thoroughly

↓

←— Powder *hamei* and dry bark
Albizia myriphylla

↓

Form paste by adding 1:1 boiled water

Form round cake

↓

Spread on bamboo basket with thick layers of husk or straw

↓

Incubate for 2–3 d

↓

Spread in open air for drying

↓

HAMEI

Figure 8.8 Indigenous method of *hamei* preparation in Manipur.

8.1.4.2 Economy

Women sell *hamei* in local markets in Manipur. The price of each piece of *hamei* is 50 paisa to Rs. 4, depending on the size of the cakes.

8.1.4.3 Microorganisms

Molds: *Mucor* spp. and *Rhizopus* spp.; yeasts: *Saccharomyces cerevisiae, Pichia anomala, P. guilliermondii, P. fabianii, Trichosporon* sp., *Candida tropicalis, C. parapsilosis, C. montana,* and *Torulaspora delbrueckii*; LAB:

Pediococcus pentosaceus and *Lactobacillus brevis* (Tamang et al. 2007a; Jeyaram et al. 2008b).

8.1.5   Khekhrii

Khekhrii is a unique starter culture different from other Himalayan starter cultures. It is prepared from fermented germinated rice in Nagaland to prepare an ethnic alcoholic drink called *zutho* or *zhuchu* by Mao tribes in Nagaland (Mao 1998).

8.1.5.1   Indigenous knowledge of preparation

Water used for making *khekhrii* is traditionally brought in a gourd (*Lagenaria siceraria* [Molina] Standley) (family Cucurbitaceae) shell from a spring (Mao 1998). Mao tribes believe that if water is brought in any other jar it may spoil the starter culture. Rice is collected and cleaned and put into an earthen jar that is filled with water brought in the gourd shell. Two pieces of charcoal and two fresh twigs of *Elscholtzia blanda* Benth. (family Lamiaceae) are also put into the jar. *Khekhrii* makers believe that the addition of charcoal pieces and *Elscholtzia blanda* are important, acting as antimicrobial regulators to keep fermenting rice from contamination. The mouth of the jar is closed tightly with fresh leaves of *Justicia adhatoda* Nees (family Acanthaceae), and the jar is kept in a warm place for 7–14 days, depending on the room temperature. In general, it takes about a week in summer and two weeks in winter. At the end of fermentation, a typical flavor comes out when the mouth of the jar is opened. After fermentation, the contents of the jar are poured into a sieve and the water is discarded, and the fermented paddy is kept inside the basket. The basket is opened after germination of the paddy has taken place. The germinated paddy is then dried in the sun and stored in a dry container for use as a starter, called *khekhrii* (Figure 8.9), which is pounded into powder and used in the preparation of *zutho* or *zhuchu*, a traditional alcoholic beverage of Nagaland.

8.1.5.2   Microorganisms

Unknown.

8.2   Alcoholic food beverages

8.2.1   Kodo ko jaanr or chyang

Kodo ko jaanr or *chyang* is the most common fermented alcoholic beverage prepared from dry seeds of finger millet (*Eleusine coracana*), locally called *kodo* in Darjeeling hills, Nepal, and Sikkim (Tamang et al. 1996). *Jaanr* is

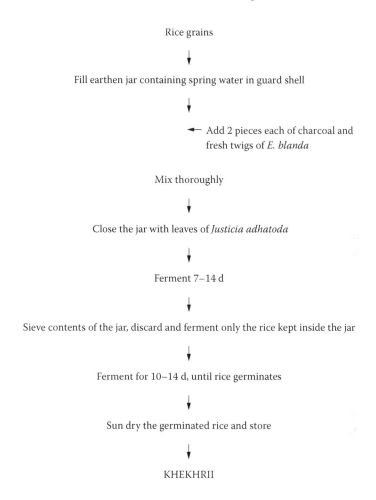

Rice grains

↓

Fill earthen jar containing spring water in guard shell

↓

← Add 2 pieces each of charcoal and
fresh twigs of *E. blanda*

Mix thoroughly

↓

Close the jar with leaves of *Justicia adhatoda*

↓

Ferment 7–14 d

↓

Sieve contents of the jar, discard and ferment only the rice kept inside the jar

↓

Ferment for 10–14 d, until rice germinates

↓

Sun dry the germinated rice and store

↓

KHEKHRII

Figure 8.9 Indigenous method of *khekhrii* preparation in Nagaland.

a common name for all alcoholic beverages in Nepali. Different ethnic groups have a name for it in their own dialect, such as *mandokpenaa thee* in Limboo, *sampicha ummaak* in Rai, *naarr paa* in Gurung, *saangla chi* in Tamang, *chirs shyaabu* in Sunwar, *paadaare haan* in Magar, *gyaar chyang* in Sherpa, *minchaa chyang* in Bhutia and Tibetan, and *mong chee* in Lepcha.

Kodo ko jaanr, prepared from dry seeds of finger millets, is one of the most popular alcoholic beverages consumed by the mountain people of the Himalayan regions in Tibet, Ladakh, Nepal, Sikkim, Darjeeling hills, and Bhutan (Tamang et al. 1996; Tamang 2005a). The custom of alcohol-drinking in the Sikkim Himalayas has been described in some historical documents (Hooker 1854; O'Malley 1907; Risley 1928; Gorer 1938).

Figure 8.10 A Tibetan lady making *kodo ko jaanr* or *chyang*.

8.2.1.1   *Indigenous knowledge of preparation*

Seeds of finger millet are cleaned, washed, and cooked for about 30 min in an open cooker. Excess water is drained off, and cooked millets are spread on a bamboo mat, locally called *mandro*, for cooling. About 1%–2% of powdered *marcha* is sprinkled over the cooked seeds, mixed thoroughly, and packed in a bamboo basket lined with fresh fern, locally called *thadre uneu* (*Thelypteris erubescens*), or banana leaves (Figure 8.10). It is then covered with sackcloth, and kept for 2–4 days at room temperature for saccharification. During the process called saccharification, a sweet aroma is emitted, and the saccharified mass is transferred into an earthen pot or into a specially made bamboo basket, called a *septu*. This is made airtight, and the seeds are fermented for 3–4 days during summer and 5–7 days in winter at room temperature (Figure 8.11).

Good-quality *jaanr* has a sweet taste with a mild alcoholic flavor. Prolonged fermentation makes the product bitter in taste and more alcoholic. Consumers generally reject *jaanr* with sour taste and unpleasant flavor.

8.2.1.2   *Mode of drinking*

Daily consumption of *kodo ko jaanr* in rural Sikkim is 101 g per capita (Tamang et al. 2007b). *Kodo ko jaanr* is consumed in a unique way in the Himalayas. About 200–500 g of *kodo ko jaanr* is put into a vessel, called a *toongbaa*, and lukewarm water is added up to its edge (Figure 8.12). After 10–15 min, the milky white extract of *jaanr* is sipped through a narrow

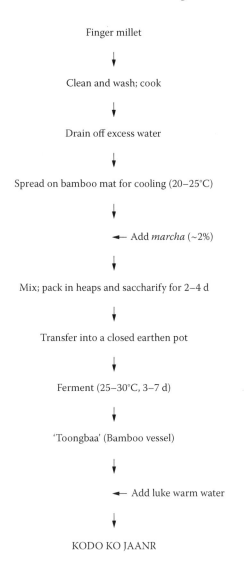

Finger millet

↓

Clean and wash; cook

↓

Drain off excess water

↓

Spread on bamboo mat for cooling (20–25°C)

↓

←— Add *marcha* (~2%)

↓

Mix; pack in heaps and saccharify for 2–4 d

↓

Transfer into a closed earthen pot

↓

Ferment (25–30°C, 3–7 d)

↓

'Toongbaa' (Bamboo vessel)

↓

←— Add luke warm water

↓

KODO KO JAANR

Figure 8.11 Indigenous method of *kodo ko jaanr* preparation in Darjeeling hills and Sikkim.

bamboo straw, called a *pipsing*, which has a hole in a side near the bottom to avoid passing of grits. Water can be added twice or thrice after sipping of the extract (Figure 8.13). Consumption of fermented finger-millet beverages in an exclusively decorated bamboo or wooden vessel (a *toongbaa*) is uncommon among other ethnic communities in Asia or elsewhere. This type of alcohol consumption or drinking is restricted to the Himalayan regions. Guests are served with *toongbaa* along with fried meat or pickle.

Figure 8.12 *Toongbaa* with *kodo ko jaanr.*

Alternatively, the thick milky white liquid pressed from the *kodo ko jaanr* is filtered under pressure using a filter called a *chhapani*. Such liquor is believed to be a good tonic for ailments and postnatal care. *Kodo ko jaanr* liquor is believed to be a good tonic for ailing persons and postnatal women. After consumption, the residual or grits of *kodo ko jaanr* are used as fodder for pigs and cattle. This is a good example of total utilization of substrate as food and fodder, as the discarded grits contain nutrients that can be used as an animal feed. Some people are economically dependent on *chyang*. *Kodo ko jaanr* is commonly available in liquor shops (locally called *gaddi*), restaurants, and hotels.

8.2.1.3   Traditional equipment used

A *septu* is a container made up of bamboo strips or wood to store *jaanr*. It is commonly used in marriages. *Chhapani* is a filter made of bamboo

Figure 8.13 A Lepcha man sipping *chee*.

strips used to filter the fermented mass. A *mandro* is a mat made of bamboo that is used to cool the cooked substrates before fermentation. The size of a *mandro* varies from 1 to 2 m². *Toongbaa* is the Limboo word (*toong* = cut properly; *baa* = bamboo) meaning a bamboo vessel cut properly; it is also called *dhungro*. A *toongbaa* is a vessel in which fermented millet seeds are placed and warm water is added. The extract is sipped through a narrow bamboo straw (a *pipsing*), which is also derived from the Limboo word *pipmasing* (*pipma* = to sip; *sing* = narrow bamboo straw). *Toongbaa* is made up of wood or bamboo or earthenware. Usually, a wooden *toongbaa* is decorated with silver lining and is provided with a lid. A *pipsing* is a narrow bamboo straw with a hole on opposite sides at one end of the straw to avoid passing of grits during sipping of *jaanr* from a *toongbaa*.

8.2.1.4   Microorganisms
Filamentous molds: *Mucor circinelloides* and *Rhizopus chinensis*; yeasts: *Pichia anomala, Saccharomyces cerevisiae, Candida glabrata,* and *Saccharomycopsis fibuligera*; LAB: *Pediococcus pentosaceus* and *Lactobacillus bifermentans* (Thapa 2001; Thapa and Tamang 2004).

8.2.2   Chyang

Chyang or *chhang* or *lugri* is prepared from a huskless local variety of barley (*Hordeum nulum*) called *sherokh* in Ladakh (Bhatia et al. 1977). *Chyang* or *lugri*, a barley-based fermented beverage, is a mildly alcoholic, thick, translucent, foamy drink with a sweet-sour taste and somewhat aromatic flavor (Batra and Millner 1976; Batra 1986).

8.2.2.1   Indigenous knowledge of preparation

Barley grains are cooked over a slow fire in water just sufficient for absorption and spread on a burlap mat or blanket to remove the free water. The cooked grains at lukewarm stage are inoculated with *phab* using 1g/kg of substrate. The dry *phab* is finely ground, diluted with *sattu* powder (a form of wheat meal) in the ratio 1:3, and sprinkled over the cooked barley and mixed. The contents are filled in drill bags, often in 20-kg batches, and tightly tied. These are further wrapped by gunny bags or woolen rags to maintain temperature around 30°C–35°C and left to ferment for 7–8 days. During this period, alcohol is produced along with some acids, and completion of fermentation is judged by a characteristic smell and the oozing of free liquid from the bag. The mass is transferred to a narrow-mouth stone or metal jar and filled up to the brim. The vessel is plugged tightly by cloth and covered with mud plaster to provide an airtight seal. This sock, as such, can be kept preserved up to 1 month in summers and 6 months in winters without adverse effects on quality (Bhatia et al. 1977).

In the traditional method of preparation of *chyang* or *lugri* in Tibet, during late March through May, high-quality grains from the previous year are soaked overnight, dewatered, spread in gunny sacks, incubated for 2–5 days in a warm place, and allowed to dry gradually. The grain is further air dried in the sun, then coarsely ground and mashed. The mash is boiled, cooled, mixed with unmalted crushed grain and fermented for 3–6 days in a cool place. The starter inoculum comes either from the unmalted grain and flowers of diverse plants that are added, or from the portion of beer added to the mash from a previous batch.

8.2.2.2   Mode of drinking

The mode of drinking is the same as for *kodo ko jaanr* or *chyang* of Sikkim. It is consumed without additional carbonation and is usually neither aged nor filtered. The *chiang* of southwestern Tibet, along the Nepal border, is sour and aged during storage for one month. *Chiang* is consumed in this area with the addition of 2–3 g of yak butter, which floats on the top of the beverage (Batra and Millner 1976).

There are several lesser-known ethnic alcoholic beverages prepared and consumed by the Tibetans in Tibet and the Bhutia in Ladakh, Himachal Pradesh, Uttarakhand, and Arunachal Pradesh (Bhatia et al. 1977; Das and Pandey 2007). These include *sing sing, shull shull, buza,* etc., prepared from barley and eaten as food beverages by the Tibetans.

8.2.2.3 Microorganisms
Yeasts: *Saccharomyces cerevisiae* and *S. uvarum* (Batra 1986).

8.2.3 Bhaati jaanr

Bhaati jaanr is an inexpensive, high-calorie, mildly alcoholic beverage that is prepared from steamed glutinous rice and that is consumed as a staple food beverage in the eastern Himalayan regions of Nepal, India, and Bhutan. *Bhaati jaanr* is a Nepali word for a fermented rice beverage that has several vernacular names: *tak thee* in Limboo, *kok umaak* in Rai, *kaiyan paa* in Gurung, *kaan chi* in Tamang, *kameshyaabu* in Sunwar, *chho haan* in Mangar, *ja thon* in Newar, *dacchhang* in Sherpa, *laayakaa chyang* in Tibetan, and *jo chee* in Lepcha. It is a traditional diet in villages for new mothers, who believe that it helps them to regain their strength (Figure 8.14).

8.2.3.1 Indigenous knowledge of preparation
During production of *bhaati jaanr*, glutinous rice is cooked for about 15 min in an open cooker, excess water is drained off, and cooked rice is spread on a bamboo mat for cooling ($\approx 40°C$). About 2%–4% of powdered *marcha* is sprinkled over the cooked rice, which is then mixed well and kept in a vessel or an earthen pot for 1–2 days at room temperature for saccharification. During saccharification, a sweet aroma is emit-

Figure 8.14 Nonfermented rice and *bhaati jaanr.*

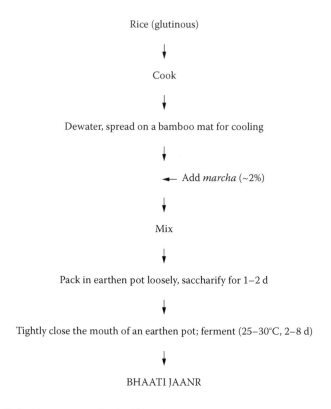

Rice (glutinous)

↓

Cook

↓

Dewater, spread on a bamboo mat for cooling

↓

←— Add *marcha* (~2%)

↓

Mix

↓

Pack in earthen pot loosely, saccharify for 1–2 d

↓

Tightly close the mouth of an earthen pot; ferment (25–30°C, 2–8 d)

↓

BHAATI JAANR

Figure 8.15 Indigenous method of *bhaati jaanr* preparation in Sikkim.

ted. After saccharification, the vessel is made airtight and fermented for 2–3 days in summer and 7–8 days in winter (Figure 8.15).

8.2.3.2   Mode of consumption

Bhaati jaanr is made into a thick paste by stirring the fermented mass with the help of a hand-driven wooden or bamboo stirrer. It is consumed directly as a food beverage. Occasionally, *bhaati jaanr* is stored in an earthenware crock for 6–9 days, and a thick, yellowish-white supernatant liquor, locally called *nigaar*, is collected at the bottom of the earthenware crock. *Nigaar* is drunk directly with or without addition of water. *Bhaati jaanr* is an inexpensive high-calorie staple food-beverage for postnatal women and ailing old persons in the villages who believe that it helps to regain their strength.

Bhaati jaanr is not sold in the local markets. People prefer to make it at home for consumption.

8.2.3.3   Similar products

Bhaati jaanr is similar to other fermented glutinous-rice beverages of Asia such as *tapé ketan* of Indonesia (Ko 1972), *lao-chao* of the Cantonese in China (Wang and Hesseltine 1970; Wei and Jong 1983), *tien-chiu-niang* of the Mandarins in China (Haard et al. 1999), *takju* of Korea (Park et al. 1977), and *khao-maak* of Thailand (Phittankpol et al. 1995).

8.2.3.4   Microorganisms

Filamentous molds: *Mucor circinelloides, M. hiemalis, Rhizopus chinensis,* and *R. stolonifer* var. *lyococcus;* yeasts: *Saccharomycopsis fibuligera, Saccharomycopsis capsularis, Pichia anomala, P. burtonii, Saccharomyces cerevisiae, S. bayanus,* and *Candida glabrata;* and LAB: *Pediococcus pentosaceus, Lactobacillus bifermentans,* and *Lb. brevis* (Tamang and Thapa 2006).

8.2.4   Poko

Poko is an alcoholic beverage prepared from rice in Nepal (Shrestha et al. 2002), similar to *bhaati jaanr* of Sikkim. It is widely used during celebrations, festivals, and ceremonies, especially by the rural population. People believe that *poko* promotes good health and nourishes the body, giving good vigor and stamina.

8.2.4.1   Indigenous knowledge of preparation

In the traditional method of *poko* production, rice is soaked overnight, steamed until cooked and sticky, and then spread to cool on a clean floor at room temperature. Powdered *marcha* or *manapu* is sprinkled on the cooked rice and mixed well; the mixture is then packed in earthen vessels. The vessel mouths are closed with leaves and straw and tied with muslin cloth. The vessels are then covered with blankets or cloths to keep them warm above a straw mat, and the rice is allowed to ferment at room temperature for 2–3 days in the summer and 4–5 days in the winter (Figure 8.16). The product is mixed every day during the fermentation period. The sticky rice is transformed to a creamy white, soft, juicy mass with a sweet, sour, mildly alcoholic, and aromatic flavor, at which point it is ready for consumption (Shrestha et al. 2002).

The mode of consumption of *poko* is same as for *bhaati jaanr*.

8.2.4.2   Microorganisms

Mold: *Rhizopus* spp.; yeasts: *Saccharomyces cerevisiae* and *Candida versatilis;* LAB: *Pediococcus pentosaceus* (Shrestha et al. 2002).

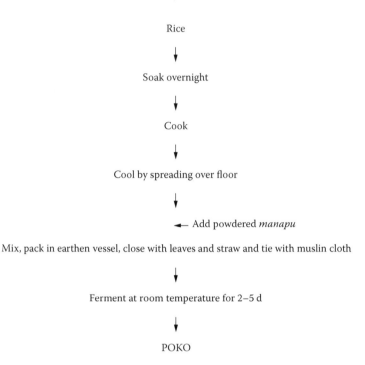

Rice

↓

Soak overnight

↓

Cook

↓

Cool by spreading over floor

↓

◄— Add powdered *manapu*

Mix, pack in earthen vessel, close with leaves and straw and tie with muslin cloth

↓

Ferment at room temperature for 2–5 d

↓

POKO

Figure 8.16 Indigenous method of *poko* preparation in Nepal.

8.2.5   *Makai ko jaanr*

Makai ko jaanr is a viscous, slightly bitter, mildly alcoholic beverage fermented from maize. Preparation and consumption of *makai ko jaanr* are confined to a few places in Nepal, Darjeeling hills, and Sikkim. The common name for maize is *makai* in the Nepali language, and the fermented product is known by different names: *makai thee* by the Limboo, *yobbhacha umaak* by the Rai, *makhain paa* by the Gurung, *maagnila jheen* by the Tamang, *aakan shyaabu* by the Sunwar, *makai haan* by the Magar, *kahni thon* by the Newar, *lichee chyang* by the Sherpa, *kinya chyang* by the Bhutia, and *kanchung chee* by the Lepcha.

8.2.5.1   *Indigenous knowledge of preparation*

Dry seeds of maize are ground and dehusked. The larger ground granules of maize, called *chekhla*, are selected for preparation of *makai ko jaanr*. *Chekhla* are washed, cooked to a thick porridge, cooled, and inoculated with powdered *marcha* (1.0%–2.0%). Saccharification and fermentation methods of *makai ko jaanr* are same as for *bhaati jaanr* (Figure 8.17).

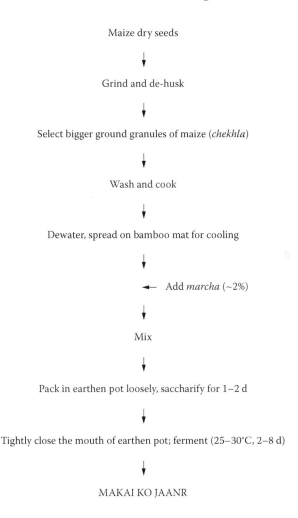

Maize dry seeds

↓

Grind and de-husk

↓

Select bigger ground granules of maize (*chekhla*)

↓

Wash and cook

↓

Dewater, spread on bamboo mat for cooling

↓

← Add *marcha* (~2%)

↓

Mix

↓

Pack in earthen pot loosely, saccharify for 1–2 d

↓

Tightly close the mouth of earthen pot; ferment (25–30°C, 2–8 d)

↓

MAKAI KO JAANR

Figure 8.17 Indigenous method of *makai ko jaanr* preparation in Kalimpong.

8.2.5.2   *Mode of consumption*

Fermented porridge is mashed, to which a filtered and desirable amount of lukewarm water is added. This extract of *makai ko jaanr* is drunk directly. It is a slightly bitter, mildly alcoholic beverage.

8.2.5.3   *Microorganisms*

Filamentous molds: *Mucor circinelloides, M. hiemalis, Rhizopus chinensis,* and *R. stolonifer* var. *lyococcus*; yeasts: *Saccharomycopsis fibuligera, Pichia anomala, P. burtonii, Saccharomyces cerevisiae, Candida glabrata*; LAB: *Pediococcus pentosaceus, Lactobacillus bifermentans,* and *Lb. brevis* (Thapa 2001).

8.2.6   Gahoon ko jaanr

Gahoon ko jaanr is an alcoholic beverage prepared from wheat (*Triticum aestivum* L.).

8.2.6.1   Indigenous knowledge of preparation
The method of preparation of *gahoon ko jaanr* is the same as that used for *kodo ko jaanr*.

8.2.6.2   Mode of drinking
The beverage is drunk directly after filtering the fermented grits. Sometimes, *gahoon ko jaanr* is mixed with *kodo ko jaanr* and filled up in *toongbaa* and consumed. *Gahoon ko jaanr* is mostly used for distillation to get *raksi*, a clear, distilled liquor.

8.2.6.3   Microorganisms
Filamentous molds: *Mucor circinelloides, M. hiemalis, Rhizopus chinensis,* and *R. stolonifer* var. *lyococcus*; yeasts: *Saccharomycopsis fibuligera, Pichia anomala, P. burtonii, Saccharomyces cerevisiae,* and *Candida glabrata*; LAB: *Pediococcus pentosaceus* and *Lactobacillus bifermentans* (Thapa 2001).

8.2.7   Atingba

Atingba is a traditional fermented beverage prepared from rice in Manipur. It is consumed by the Meitei as food beverage.

8.2.7.1   Indigenous knowledge of preparation
In *atingba* preparation, rice is cooked and cooled to room temperature. *Hamei* is crushed into powder and mixed with the cooked rice at the rate of 5 cakes/10 kg. The mixture is kept inside earthen pots covered with leaves of *hangla* (*Alocasia* sp.) and fermented for 3–4 days in summer and 6–7 days in winter. This is followed by 2–3 days submerged fermentation in earthen pots to get the fermented beverage called *atingba* (Figure 8.18). It is distilled to yield a clear-liquor alcoholic drink called *yu* in Manipur.

8.2.7.2   Mode of consumption
Atingba is consumed as it is or distilled into a clear liquor called *yu*.

8.2.7.3   Microorganisms
Same as *hamei*.

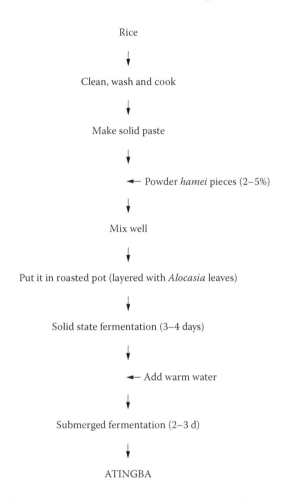

Rice

↓

Clean, wash and cook

↓

Make solid paste

↓

← Powder *hamei* pieces (2–5%)

↓

Mix well

↓

Put it in roasted pot (layered with *Alocasia* leaves)

↓

Solid state fermentation (3–4 days)

↓

← Add warm water

↓

Submerged fermentation (2–3 d)

↓

ATINGBA

Figure 8.18 Indigenous method of *atingba* preparation in Manipur.

8.2.8   *Zutho or zhuchu*

Zutho or *zhuchu* is an ethnic alcoholic beverage of Nagaland prepared from rice by the Mao Naga.

8.2.8.1   *Indigenous knowledge of preparation*

Different types of rice are used for the preparation of *zutho* or *zhuchu* (Mao 1998). However, the product is obtained from the glutinous type of rice widely cultivated by the Naga. The rice is soaked overnight and then washed thoroughly, draining off the excess water. The soaked rice is then pounded into flour with a mortar, put in a big bamboo bucket, and mixed with boiling water by stirring with a long bamboo spoon. It is then left to cool down. When it completely cools down, *khekhrii* powder is stirred and left for about 6–8 h for brewing, after which the whole mixture is poured into a big earthen jar. More water is added to make the volume up to the neck and kept for fermentation for 3–4 days, during which it forms a profuse whitish froth, to get the final product called *zutho* or *zhuchu*. It has a sweet taste with an acid flavor. The quality of *zutho* remains acceptable for up to 4–5 days, after which it becomes slightly sour and loses its good taste and flavor.

8.2.8.2   Mode of drinking

Zutho is a common alcoholic beverage in Nagaland. A similar alcoholic beverage called *nchiangne* is prepared from red rice in Nagaland.

8.2.8.3   Microorganisms

Yeasts: *Saccharomyces cerevisiae* (Teramoto et al. 2002).

8.2.9   Sura

Sura, an ethnic alcoholic beverage of Kullu of Himachal Pradesh, is fermented from millet, locally called *kodra* or *kached*, having long storage life (Thakur et al. 2004). No specific starter or inoculum is used for *sura* preparation. It is a natural fermentation process in which native microflora carry out starch hydrolyzing and ethanol formation.

8.2.9.1   Indigenous knowledge of preparation

No specific starter or inoculum is used for *sura* preparation. Sometimes, a starter called *dhehli* is also added during production of *sura*. *Dhehli* preparation is an annual community effort, in which village elders collect approximately 36 wild herbs from the Himalayan forests on the 20th day of the *Bhadrapada* (religious) month, which usually falls on 5 or 6 September. On the next day, all 36 wild herbs are crushed in a *ukhal* (stone with a large conical cavity) using a *mussal* (a wooden pestle), and the extract as well as the plant biomass are added into the *sattu* (flour of roasted barley) and are roughly kneaded. This mash is put into a brick-shaped wooden mold and then sun dried. The resultant brick, locally called a *dhehli* (Figure 8.19), is divided among the villagers to be used whenever *sura* is to be prepared. *Dhehli* provides bioactive compounds as well as a stimulatory effect (Thakur et al. 2004).

8.2.9.2   Mode of drinking

Sura is consumed during local festivals (*shoeri saja*) and marriage ceremonies in Himachal Pradesh.

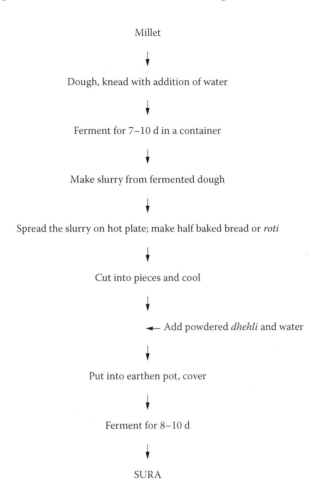

Millet

↓

Dough, knead with addition of water

↓

Ferment for 7–10 d in a container

↓

Make slurry from fermented dough

↓

Spread the slurry on hot plate; make half baked bread or *roti*

↓

Cut into pieces and cool

↓

◀— Add powdered *dhehli* and water

↓

Put into earthen pot, cover

↓

Ferment for 8–10 d

↓

SURA

Figure 8.19 Indigenous method of *sura* preparation in Himachal Pradesh.

8.2.9.3   Microorganisms

Unknown.

8.3   Distilled liquor or alcoholic drink

8.3.1   Raksi

Raksi is an ethnic Himalayan alcoholic drink with a characteristic aroma, distilled from the traditional fermented cereal beverages such as *kodo ko jaanr/chyang, bhaati jaanr, makai ko jaanr, gahoon ko jaanr,* etc. (Kozaki et al. 2000). Fermented masses of buckwheat, potato, canna, and cassava roots are also distilled to get *raksi*. *Raksi* is a common term in Nepali meaning alcoholic drink. *Raksi* has several vernacular names, including *aarak* (Tibetans, Bhutia, Drukpa, Sherpa), *aarok* (Lepcha), *aaerak* (Tamang), *aayala* (Newar), *sijongwaa aara* (Limboo), *aarakha/hemma* (Rai), *paa* (Gurung), *rindho* (Sunwar), *dhise* (Magar), etc.

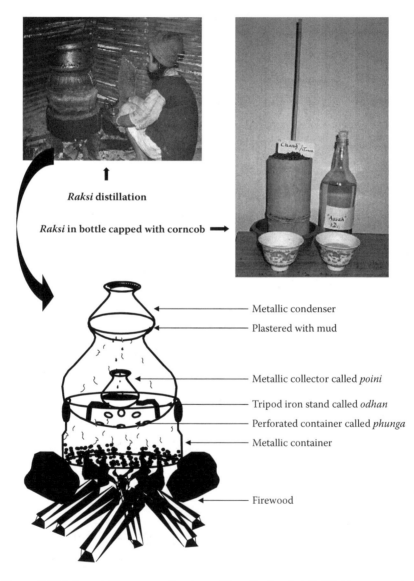

Raksi distillation

Raksi in bottle capped with corncob ➡

— Metallic condenser
— Plastered with mud

— Metallic collector called *poini*
— Tripod iron stand called *odhan*
— Perforated container called *phunga*
— Metallic container

— Firewood

Figure 8.20 Internal diagrammatic view of the ethnic *raksi* distillation apparatus.

8.3.1.1   Indigenous knowledge of alcohol distillation

Traditional fermented beverages prepared from cereals such as *bhaati jaanr, makai ko jaanr, kodo ko jaanr, gahoon ko jaanr,* etc., are distilled in a large cylindrical metallic vessel measuring 40 × 30 × 25 cm for 2–3 h continuously over firewood in an earthen oven. Above the main cylindrical vessel, a perforated container called a *phunga* is placed, inside which a small metallic collector, locally called *poini,* is kept on an iron tripod, locally called *odhan,* to collect the distillate (*raksi*). Another metallic vessel with cold water is placed above the *phunga* as condenser. The bottom of the condenser vessel is plastered by mud with the tip of the *phunga* to prevent excess ventilation during distillation (Figure 8.20). Water is replaced three to five times after it is heated. Condensed *raksi* is collected in a small metallic vessel (*poini*). *Raksi* prepared after replacing the condensing water three times is known as *theen pani raksi;* this contains a high amount of alcohol and is traditionally prepared for religious purposes. *Raksi* prepared after replacing the condensing water five times is known as *panch pani raksi,* which is a common alcoholic drink. The traditional distillation apparatus can distill 2–4 kg of *jaanr* to get 1–2 liters of *raksi* after replacing condensing water thrice. *Raksi* is usually stored in bottles capped with a piece of dry corncob. Sometimes, petals of *Rhododendron* spp. are mixed during distillation to give a distinct aroma to *raksi.* This type of *raksi* is commonly prepared in Rhododendron-growing regions of the Himalayas (Tamang et al. 1996). The most common fermented beverages used for traditional distillation are *makai ko jaanr,* prepared from maize, and *bhaati jaanr,* prepared from rice, in the Himalayas.

8.3.1.2   Mode of drinking

Raksi is drunk directly without addition of water along with fried meat or any other curry. It is traditionally drunk by the *matwali* (alcohol drinkers), including non-Brahmin Nepali, Bhutia, Tibetans, Lepcha, Drukpa, etc. Sometimes, *raksi* is made to sizzle in a hot vessel containing butter. The vessel is covered with a lid as long as it emits a sizzling sound. Such *raksi* is called *jhane ko raksi,* which is drunk hot as a relief from cough and cold.

8.3.1.3   Economy

Some people are economically dependent on *raksi,* which is commonly available in liquor shops, restaurants, and hotels. *Raksi* costs about Rs. 2–3 per 100 ml in Darjeeling hills and Sikkim.

8.3.1.4   Distilled alcoholic drinks similar to raksi

Similar distilled alcoholic drinks prepared from fermented cereal products in North East India are *yu* of Manipur distilled from *atingba, saké* of

Japan (Inoue et al. 1992), *soju* of Korea (Lee and Kim 1993), and *krachae* or *nam-khaao* or *sato* of Thailand (Vachanavinich et al. 1994).

8.4 Microbiology

8.4.1 Starter culture

The Himalayan ethnic starter culture is a consortium of microorganisms consisting of mycelial fungi, amylolytic and alcohol-producing yeasts, and lactic acid bacteria. *Marcha* and *hamei* or similar traditional starter-making technologies reflect the native skill of ethnic people in subculturing of desirable inocula (microorganisms) from a previous batch to a new culture using rice or wheat as a starchy base or medium. This indigenous technique of "microbiology" preserves the functional microorganisms necessary for fermentation of starchy substrates to alcoholic beverages in the Himalayas. These types of starter cultures retain their potencies in situ for two years or more.

Mold population in *marcha* has been detected at the level of 10^6 cfu/g, whereas the loads of yeasts and LAB were 10^8 cfu/g and 10^7 cfu/g, respectively (Tamang 1992). The first preliminary study of microbiology of *marcha* samples collected from Sikkim was carried out by Kobayashi et al. (1961), who reported the presence of *Rhizopus oryzae*, *Mucor praini*, and *Absidia lichtheimi* in *marcha*. They also mentioned the fermented beverage product *kodok jar*, which is actually *kodo ko jaanr*. Batra and Millner (1974) reported the presence of *Hansenula anomala* var. *schneggii* (= *Pichia anomala*) in *marcha* collected from Kalimpong. Hesseltine et al. (1988) isolated *Mucor* and *Rhizopus* spp. from *marcha* samples of Nepal. Uchimura et al. (1990) reported the dominant yeast *Saccharomycopsis* and the molds *Penicillium* sp. and *Aspergillus* sp. in a Bhutanese mixed starter called *chang-poo* or *phab*. Tamang (1992) studied the microbiology of *marcha* of Darjeeling hills and Sikkim. Species of filamentous molds were identified as *Mucor circinelloides* forma *circinelloides* van Tieghem, *Mucor* sp. close to *M. hiemalis* sensu lato, *R. chinensis* Saito, and *R. stolonifer* var. *lyococcus* (Ehrenb.) Stalp & Schipper. Species of *Mucor* were found to be more prevalent than species of *Rhizopus* in *marcha* (Tamang 1992). *Aspergillus*, *Penicillium*, *Amylomyces*, and *Actinomucor* are not present in *marcha* of the Himalayas. According to Tanaka and Okazaki (1982), *Rhizopus* grows well on nonsteamed grains, but its specific growth rate is decreased remarkably by steaming rice grain. The decrease is due to the heat denaturation of proteins in rice grain. *Rhizopus* and *Mucor* have also been reported in other similar rice-based starter cultures, such as *ragi* and *chiuyueh* (Hesseltine et al. 1988), *bubod* (Tanimura et al. 1978), and *loogpang* (Pichyangkura and Kulprecha 1977). The Himalayan starter cultures show dominance of species of *Mucor* and *Rhizopus* among the mycelial fungi; species of *Aspergillus* and

Penicillium have not been reported in any Himalayan mixed starter except *mana* of Nepal, which shows the presence of *Aspergillus oryzae* (Nikkuni et al. 1996).

Based on morphological, physiological, and molecular characterization, yeasts isolated from *marcha* are first classified into four groups (Group I, II, III, IV), and are identified as *Saccharomyces bayanus* (Group I), *Candida glabrata* (Group II), *Pichia anomala* (Group III), *Saccharomycopsis fibuligera*, *Saccharomycopsis capsularis*, and *Pichia burtonii* (Group IV) (Tsuyoshi et al. 2005). Among them *Saccharomyces bayanus*, *Candida glabrata*, and *Pichia anomala* strains produced ethanol, whereas the isolates of *Saccharomycopsis fibuligera*, *Saccharomycopsis capsularis,* and *Pichia burtonii* have high amylolytic activity. Because all *marcha* samples tested contained both starch degraders and ethanol producers, it is presumed that all groups of yeasts contribute to starch-based alcohol fermentation (Tsuyoshi et al. 2005). Among them, the strains of *Saccharomyces bayanus*, *C. glabrata*, and *P. burtonii* have not been isolated from *marcha* in previous reports (Batra and Millner 1974; Hasseltine and Kurtzman 1990). The identified yeast species, except for *Saccharomyces bayanus* and *Candida glabrata*, are also encountered in and isolated from several other Asian amylolytic starters (Hadisepoetro et al. 1979; Hesseltine et al. 1988; Hesseltine and Kurtzman 1990; Deak 1991). Although the species of *Saccharomyces bayanus* have not been isolated from any other Asian amylolytic starters, the closely related species of *Saccharomyces cerevisiae* has been isolated from *ragi* and *banh men* (Hesseltine et al. 1988; Lee and Fujio 1999). *Saccharomycopsis fibuligera* is the most dominant yeast in *marcha* (Tamang and Sarkar 1995), which is typically found growing on cereal products (Hesseltine and Kurtzman 1990). The most frequent yeast species present in *hamei* is *Pichia anomala* (41.7%), followed by *S. cerevisiae* (32.5%) and *Trichosporon* sp. (8%); the identity of major groups was confirmed by additional restriction digestion of ITS (internal transcribed spacer) region with Hind III, EcoRI, Dde I, and Msp I (Jeyaram et al. 2008b).

Among LAB species, *P. pentosaceus*, *Lb. bifermentans*, and *Lb. brevis* are present in *marcha* (Hesseltine and Ray 1988; Tamang and Sarkar 1995; Tamang et al. 2007a). The RAPD-PCR (randomly amplified polymorphic DNA–polymerase chain reaction) characterization of LAB strains revealed the close relationship of the two *Pediococcus* isolates from *hamei* (Tamang et al. 2007a). The application of the rep PCR with GTG_5 primer confirmed the homofermentative *Lactobacillus* HB:B1 as *Lb. plantarum*, and the species-specific PCR confirmed the heterofermentative *Lactobacillus* MA:R5 as *Lb. brevis. Pediococcus pentosaceus* and *Lb. plantarum* are present in *hamei*. Pediococci are more predominant lactics than lactobacilli in both *hamei* and *marcha* (Tamang et al. 2007a).

LAB strains isolated from *hamei* show stronger antimicrobial activities than LAB from *marcha* (Tamang et al. 2007a). Using culture supernatants, only *P. pentosaceus* HS:B1 (*hamei*) has been found to produce bacteriocin against *Listeria innocua* DSM 20649 and *Listeria monocytogenes* DSM 20600 at the level of 128 AU/ml and 32 AU/ml, respectively (Tamang et al. 2007a). This suggests that antimicrobial properties of functional LAB can reduce the number of other undesired microorganisms in this traditional starter. None of the strains of LAB produces biogenic amines, which is a good indicator for starter cultures. It has been observed that yeast strains— mostly *Saccharomyces cerevisiae, Pichia anomala,* and *Saccharomycopsis fibuligera*—produce a high amount of alcohol, comparable to that of LAB strains, whose presence is negligible. This confirms the strains of LAB present in *hamei* and *marcha* have little or no role in alcohol production (Tamang et al. 2007a). All LAB strains show strong peptidase, arylamidase, and glucosaminidase activities. However, they show no detectable esterase, lipase, phosphatase, and protease activity. None of the LAB show amylolytic activity, indicating that they have no role in the saccharification and liquefaction of starchy substrates. The inability to utilize starch by LAB indicates that they are not significant contributors to the breakdown of starch of substrates during preparation of *hamei* and *marcha* itself or any beverage. It is concluded that LAB present in *hamei* and *marcha* play a role in imparting flavor, antagonism of pathogens, and acidification of the substrates (Tamang et al. 2007a). The role of LAB in the Asian starter cultures is likely to give flavor to the product with a pleasant taste (Hesseltine 1983). Isolates of *marcha* show relatively weak esterase and lipase activities (except by *Pichia anomala*) as compared with phosphatase activities. Yeast strains, mainly *Saccharomycopsis fibuligera* and *Pichia anomala,* show strong phosphatase and peptidase (leucine arylamidase) activities.

Preliminary screenings of amylolytic activities of all marcha isolates were tested in starch agar plates (Thapa 2001). On the basis of amylolytic activity, four strains of *Rhizopus* spp., two strains of *Mucor* spp., five strains of *Saccharomycopsis fibuligera,* four strains of *Pichia anomala,* four strains of *Saccharomyces cerevisiae,* and three strains of *Candida glabrata* are selected for liquefying and saccharifying activities. None of the LAB strains show amylolytic activity; hence, they were not selected for amylolytic enzyme assay. Saccharifying activities are mostly shown by *Rhizopus* spp. and *Saccharomycopsis fibuligera,* whereas liquefying activities are shown by *S. fibuligera* and *S. cerevisiae.* In *tapé,* a fermented rice, sweet-sour paste of Indonesia, *S. fibuligera* produces mainly α-amylase, and *Rhizopus* sp. produced glucoamylase (Suprianto et al. 1989). *Saccharomycopsis fibuligera* has major roles in amylase production, whereas *Rhizopus* seems to supplement the saccharification (Sukhumavasi et al. 1975; Cronk et al. 1977; Wei and Jong 1983; Uchimura et al. 1990, 1991; Yokotsuka 1991). *Rhizopus* is known to produce good amount of glucoamylase (Ueda and Kano 1975; Selvakumar et al. 1996). *Rhizopus* spp. and amylolytic yeasts (mostly *S. fibuligera*) degrade starch and produce glucose, and alcohol-producing yeasts (species of *Saccharomyces* and *Pichia*) rapidly grow on the resultant glucose to produce ethanol (Thapa 2001).

The average population of LAB in *hamei* is 10^7 cfu/g (Tamang et al. 2007a). A *mana* sample of Nepal contained 1.5×10^6 cfu/g of mucorales, 3.5×10^7 cfu/g of aspergilli, and 1.1×10^5 cfu/g of LAB, while yeast was present at less than 10^3 cfu/g (Nikkuni et al. 1996). *Aspergillus oryzae* is identified as a dominant organism in the *mana* sample, which had metulae with length of 7–15 μm and proved to be aflatoxin-negative (Nikkuni et al. 1996).

Yeast and lactics are present in high numbers in *manapu* starters, whereas molds are dominant in wheat-based *mana* samples (Shrestha et al. 2002). The yeast and LAB population in *manapu* is 10^5 to 10^9 cfu/g, whereas the mold population in *mana* is 10^7 cfu/g (Shrestha et al. 2002). *Saccharomyces cerevisiae*, *Candida versatilis*, *P. pentosaceus*, and *Rhizopus* spp. have been reported from *manapu* of Nepal (Shrestha et al. 2002).

8.4.2   Kodo ko jaanr

The population of yeasts in *kodo ko jaanr* was detected at the level of 10^7 cfu/g, whereas that of LAB was comparatively less ($\approx 10^5$ cfu/g) (Thapa and Tamang 2004). Filamentous molds were not recovered in any finished product of *kodo ko jaanr*, indicating that molds have roles only in the initial phase of fermentation, mostly in saccharification of the substrates. Yeasts consist of *Pichia anomala*, *Saccharomyces cerevisiae*, *Candida glabrata*, *Saccharomycopsis fibuligera*, and LAB consist of *Pediococcus pentosaceus* and *Lactobacillus bifermentans* in *kodo ko jaanr* samples. There is no occurrence of Enterobacteriaceae, *Bacillus cereus*, or *Staphylococcus aureus* in *kodo ko jaanr* (Thapa and Tamang 2004). *Marcha*, used as a starter culture, supplements all of these functional microorganisms in *kodo ko jaanr* fermentation (Thapa 2001). *Kodo ko jaanr* is a safe food beverage, since no pathogenic contaminants are detected in the product. This is mainly due to low pH and the increased acidity and alcohol content in the product. The pH, moisture, acidity, and alcohol contents of the product were 4.1, 69.7%, 0.27%, and 4.8%, respectively (Thapa and Tamang 2004).

8.4.2.1   In situ fermentation of kodo ko jaanr

Microbiological and physicochemical changes during in situ fermentation of *kodo ko jaanr* was studied every day from 0 to 10 days of incubation (Thapa and Tamang 2006). The population of filamentous molds (*Mucor circinelloides*, *Rhizopus chinensis*, and *Rhizopus stolonifer*, which were originated from *marcha*) declined significantly ($p < .05$) every day during fermentation and finally disappeared on the fifth day. The load of yeasts (*Saccharomycopsis fibuligera*, *Pichia anomala*, *Saccharomyces cerevisiae*, and *Candida glabrata*) increased significantly ($p < .05$) from 10^5 cfu/g to 10^7 cfu/g by the second day, indicating their roles in amylase production during fermentation. Subsequently, the load of LAB (*Pediococcus pentosaceus* and *Lb. bifermentans*) increased significantly ($p < .05$) from 10^6 cfu/g to 10^8 cfu/g on the first day and decreased significantly ($p < .05$) to a level of 10^3 cfu/g at the end of fermentation. Total viable counts increase significantly ($p < .05$) on the first day and gradually decreased every day during fermentation.

Previous research has shown that the temperature of fermenting finger millet increased significantly ($p < .05$) from 26°C to 30°C within 2 days due to the exponential growth of mixed populations of microorganisms, and then remained constant around 28°C. The pH decreased significantly ($p < .05$) from 6.4 to 4.1 until the second day of fermentation, after which the decline in pH was nonsignificant. Titratable acidity increased significantly ($p < .05$) until the fourth day, and remained the same until the end. The cause of the increase in acidity and the consequent drop in pH during fermentation of cereal is likely due to utilization of free sugars of the substrate by yeasts and LAB (Efiuvwevwere and Akona 1995; Zvauya et al. 1997), since all the strains were able to ferment glucose. Alcohol content increases significantly ($p < .05$) from 0.1% to 6.9% by the sixth day and slightly decreased to 6.5% at the end. The reducing-sugar content increases significantly ($p < .05$) until the third day, followed by a decrease in total sugar content. This is due to maximum breakdown of starch substrates to reducing sugars by amylolytic enzymes, which are produced by molds and yeasts during fermentation. Maximum activities of saccharification (glucoamylase) and liquefaction (α-amylase) of finger millets are observed on the second day of fermentation. Saccharifying activities are mostly shown by *Rhizopus* spp. and *Saccharomycopsis fibuligera*, whereas liquefying activities are shown by *Saccharomycopsis fibuligera* and *S. cerevisiae* (Thapa and Tamang 2006). This experiment explains that *S. fibuligera* and *Rhizopus* spp. play an important role in saccharification and liquefaction processes during in situ fermentation of *jaanr*, breaking starch of finger millets into glucose for ethanol production. *Mucor* spp., *Pichia anomala*, and *Candida glabrata* may supplement the saccharification.

8.4.2.2 Improved method of kodo ko jaanr production

For testing the ability of selected strains to produce *kodo ko jaanr*, *Rhizopus chinensis* MJ:R3 and *Saccharomycopsis fibuligera* KJ:S5 (strains previously isolated from native *marcha*) (Thapa 2001) were selected on the basis of highest saccharifying and liquefying activities, respectively (Tamang and Thapa 2006). *Mucor circinelloides* MS:M7, *Saccharomyces cerevisiae* MJ:YS2, *Candida glabrata* MS:YC5, and *Pichia anomala* MA:YP2 were also selected on the basis of high amylolytic activities among the same genera; *Lactobacillus bifermentans* MA:R5 and *Pediococcus pentosaceus* MA were selected randomly. Each selected strain was tested for its ability to produce the final product either singly or in combination. *Kodo ko jaanr* prepared by a

Table 8.1 Changes in pH, Reducing Sugar, and Alcohol in Kodo ko Jaanr Prepared by Selected Strains

Strain	pH 6.37 ± 0.01		Reducing Sugar (%) 0.4 ± 0.16		Alcohol (%) 0.1 ± 0.06	
Cooked millet (noninoculated)	2 days	6 days	2 days	6 days	2 days	6 days
Rhizopus chinensis MJ:R3 with:						
Mucor circinelloides MS:M7	4.6 ± 0.02[e]	4.8 ± 0.01[e]	2.10 ± 0.19[ef]	3.10 ± 0.24[b]	0.25 ± 0.06[f]	0.83 ± 0.06[f]
Saccharomyces cerevisiae MJ:YS2	4.3 ± 0.01[f]	4.4 ± 0.01[h]	4.21 ± 0.05[c]	3.57 ± 0.05[a]	2.50 ± 0.07[a]	4.40 ± 0.13[a]
Candida glabrata MS:YC5	4.3 ± 0.01[f]	4.3 ± 0.00[i]	3.95 ± 0.11[c]	2.00 ± 0.19[ef]	0.70 ± 0.06[d]	1.80 ± 0.00[d]
Pichia anomola MA:YP2	4.6 ± 0.00[e]	4.7 ± 0.01[f]	2.49 ± 0.08[def]	3.13 ± 0.45[b]	0.76 ± 0.00[c]	2.20 ± 0.06[c]
Lb. bifermentans MA:R5 + *P. pentosaceus* MA:C1	4.1 ± 0.01[h]	4.2 ± 0.01[i]	2.71 ± 0.46[d]	2.89 ± 0.16[b,c]	1.00 ± 0.06[b]	2.50 ± 0.06[b]
Saccharomycopsis fibuligera KJ:S5 with:						
Rhizopus chinensis MJ:R3	4.2 ± 0.02[g]	4.6 ± 0.02[g]	6.28 ± 0.14[a]	2.55 ± 0.05[cd]	0.40 ± 0.07[e]	1.00 ± 0.06[e]
Mucor circinelloides MS:M7	6.0 ± 0.01[c]	6.1 ± 0.02[b,c]	4.94 ± 0.04[b]	2.22 ± 0.05[de]	0.15 ± 0.00[g]	0.30 ± 0.00[i]
Saccharomyces cerevisiae MJ:YS2	6.1 ± 0.02[b]	6.1 ± 0.01[b]	4.74 ± 0.75[b,c]	2.63 ± 0.11[c]	0.22 ± 0.06[f]	0.50 ± 0.06[hi]
Candida glabrata MS:YC5	5.9 ± 0.00[d]	6.0 ± 0.04[d]	3.07 ± 0.20[d]	1.65 ± 0.01[f]	0.15 ± 0.00[g]	0.45 ± 0.00[i]
Pichia anomola MA:YP2	6.1 ± 0.00[b]	6.1 ± 0.00[c]	2.61 ± 0.29[de]	1.92 ± 0.11[ef]	0.15 ± 0.06[g]	0.68 ± 0.06[g]
Lb. bifermentans MA:R5 + *P. pentosaceus* MA:C1	6.0 ± 0.01[c]	6.1 ± 0.02[b,c]	2.29 ± 0.08[def]	2.64 ± 0.04[c]	0.20 ± 0.06[g]	0.75 ± 0.00[fg]
All strains*	6.4 ± 0.02[a]	6.4 ± 0.00[a]	1.91 ± 0.41[f]	1.98 ± 0.02[ef]	0.05 ± 0.06[h]	0.15 ± 0.00[k]

Note: Data represent the means ± SD of three batches of fermentation. Values bearing different superscripts in each column show statistical difference ($p < .05$).

* Cell mixture of all strains identified in this table.

Table 8.2 Sensory Evaluation of Kodo ko Jaanr Produced by Selected Strains

Strain	Aroma	Taste	Texture	Color	General acceptability
Rhizopus chinensis **MJ:R3 with:**					
Mucor circinelloides MS:M7	2.00 ± 0.93ab	1.86 ± 0.52b	2.14 ± 0.83b	2.86 ± 0.83b,c,d	2.29 ± 0.70b,c,d
Saccharomyces cerevisiae MJ:YS2	3.43 ± 0.50a	2.79 ± 0.36a	3.43 ± 0.90a	4.29 ± 1.03a	4.43 ± 0.73a
Candida glabrata MS:YC5	2.43 ± 0.90ab	1.57 ± 0.73b	2.57 ± 0.73ab	3.00 ± 0.93a,b,c	2.43 ± 0.50b,c
Pichia anomola MA:YP2	3.43 ± 0.90a	1.86 ± 0.69b	2.43 ± 0.50ab	3.71 ± 0.88ab	2.57 ± 0.50b
Lb. bifermentans MA:R5 + *P. pentosaceus* MA:C1	2.29 ± 0.88ab	1.79 ± 0.53c	2.43 ± 1.18ab	3.14 ± 0.99a,b,c	2.29 ± 0.88b,c,d
Saccharomycopsis fibuligera **KJ:S5 with:**					
Rhizopus chinensis MJ:R3	1.86 ± 0.83ab	1.29 ± 0.36b	2.43 ± 0.50ab	1.86 ± 0.64cde	1.57 ± 0.50b,c,d,f
Mucor circinelloides MS:M7	1.86 ± 0.83ab	1.21 ± 0.36b	2.29 ± 0.88ab	1.86 ± 0.83cde	1.43 ± 0.50cdef
Saccharomyces cerevisiae MJ:YS2	1.43 ± 0.73b	1.14 ± 0.35b	1.71 ± 0.88b	1.29 ± 0.70e	1.14 ± 0.35ef
Candida glabrata MS:YC5	1.43 ± 0.73b	1.21 ± 0.36b	2.00 ± 0.76b	1.29 ± 0.70e	1.29 ± 0.70def
Pichia anomala MA:YP2	2.29 ± 0.88ab	1.21 ± 0.30b	2.00 ± 1.07b	1.43 ± 0.73e	1.43 ± 0.73cdef
Lb. bifermentans MA:R5 + *P. pentosaceus* MA:C1	1.57 ± 0.73b	1.29 ± 0.36b	2.29 ± 1.03ab	1.57 ± 0.90de	1.43 ± 0.73cdef
All strains*	1.43 ± 0.73b	1.00 ± 0.00b	1.43 ± 0.73b	1.14 ± 0.35e	1.00 ± 0.00f

Note: Market *kodo ko jaanr* used as control: score 1, bad; score 5, good. Data represent the mean scores ± SD ($n = 7$). Values bearing different super-scripts in each column show statistical difference ($p < .05$).

* Cell mixture of all strains identified in this table.

combination of *Rhizopus chinensis* MJ:R3 and *Saccharomycopsis fibuligera* KJ:S5 showed significantly higher ($p < .05$) reducing-sugar contents during a saccharification period of 2 days, with a low alcohol content of 1% after 6 days (Table 8.1). After 6 days, cell-suspension mixtures of *Rhizopus chinensis* MJ:R3 and *Saccharomyces cerevisiae* KJ:S5 produced *jaanr* with significantly higher ($p < .05$) reducing sugars and a higher alcohol content of 4.4% than *jaanr* samples fermented by other strains (Thapa and Tamang 2006).

The *jaanr* produced by selected combination of strains was subjected to sensory evaluation (Table 8.2). There was no significant difference ($p < .05$) in the aroma attributes of *jaanr* prepared by a cell-suspension mixture of *Rhizopus chinensis* MJ:R3 with other strains, except *jaanr* prepared by a combination of *Saccharomycopsis fibuligera* KJ:S5 with *S. cerevisiae* MJ:YS2, *Candida glabrata* MS:YC5, and a consortium of above-mentioned strains. There was a significant difference ($p < .05$) in the taste scores of *jaanr* prepared by cell mixtures of *Rhizopus chinensis* MJ:R3 and *Saccharomyces cerevisiae* MJ:YS2 with that of other strains. However, significant differences ($p < 0.05$) in texture and color scores were observed in some *jaanr* samples. *Jaanr* prepared by a combination of *Rhizopus chinensis* MJ:R3 and *Saccharomycopsis fibuligera* KJ:S5 had a desirable sweet-sour taste but an unpleasant odor due to low alcohol content. Hence, based on sensory criteria, *jaanr* produced by these strains was unacceptable to consumers. *Kodo ko jaanr* prepared by a combination of *Rhizopus chinensis* MJ:R3 and *Saccharomyces cerevisiae* MJ:YS2 showed significantly higher ($p < .05$) scores in general acceptability. *Kodo ko jaanr* prepared by these strains had a mildly alcoholic sweet flavor, significantly acceptable ($p < .05$) to consumers. *Saccharomyces cerevisiae* has shown a strong tendency to ferment sugars into alcohol (Kozaki and Uchimura 1990). *Jaanr* prepared by a combination of *Rhizopus chinensis* MJ:R3 and *S. fibuligera* KJ:S5 has a sweet-sour taste, but due to its low-alcohol content, the product has an unpleasant odor and is rejected by the consumers. *Saccharomycopsis fibuligera* is reported to produce high biomass during fermentation of cassava starch, which leads to low ethanol yield (Reddy and Basappa 1996).

The results of the consumers-preference trial show that *kodo ko jaanr* prepared by a mixture of *Rhizopus chinensis* MJ:R3 and *Saccharomyces cerevisiae* MJ:YS2 as starter is more acceptable than the *kodo ko jaanr* prepared by *marcha*. Market *jaanr* is liked extremely by 10% of the consumers, very much by 30% and moderately by 60%. In contrast, the laboratory-made *jaanr* prepared by the selected pure culture starter was liked extremely by 40%, very much by 50%, and moderately by 10% of the consumers (Thapa and Tamang 2006). To make a good-quality *kodo ko jaanr* or *chyang*, a consortium of a selected strain of mold (*Rhizopus*) and amylolytic alcohol-producing yeast (*Saccharomyces cerevisiae*) is recommended. Similar observations while using a mixture of mold and yeast in production of Vietnamese rice wine have been reported (Dung et al. 2006b). *Kodo ko jaanr* prepared by a mixed pure culture may have more advantages over *kodo ko jaanr* or *chyang* prepared by using *marcha* or *phab* due to better product quality, better product consistency, and maximum utilization of substrates.

8.4.3   Bhaati jaanr

Microbiological analysis of *bhaati jaanr* shows the yeast population at the level of 10^7 cfu/g, whereas that of the lactic acid bacterial load is found at the range of 10^4–10^6 cfu/g (Thapa 2001). Yeast populations have been found to be higher than that of LAB in *bhaati jaanr* samples, and filamentous molds were absent in the final product (Thapa 2001).

8.4.3.1   In situ fermentation of bhaati jaanr

Bhaati jaanr is prepared using a native *marcha*, following the traditional method (Tamang and Thapa 2006). Growth kinetics during fermentation were carried out daily within a range of 0–10 days. Filamentous molds disappeared after the fifth day. Yeasts increased significantly ($p < .05$) from 10^5 cfu/g to 10^8 cfu/g by the second day. LAB increased significantly ($p < .05$) from 10^6 cfu/g to 10^7 cfu/g on the first day and decreased significantly ($p < .05$) to 10^5 cfu/g at the end of the fermentation. The pH decreased and acidity increased during fermentation. Alcohol content increased significantly ($p < .05$) up to 10% on the tenth day. Reducing-sugar content increased significantly ($p < .05$) until the third day, followed by a decrease in total sugar content. Total sugar contents decreased significantly ($p < .05$) throughout the fermentation. This is due to maximum starch breakdown of substrates to reducing sugars by amylolytic enzymes produced by molds and yeasts during fermentation (Nout and Aidoo 2002). Maximum activities of saccharification and liquefaction of rice were observed on the third day of fermentation. This result suggests that yeasts, probably *Saccharomycopsis fibuligera* and filamentous molds, contribute in saccharification and liquefaction of glutinous rice, breaking the starch of substrates into glucose for alcohol production, and also in aroma formation in *bhaati jaanr* preparation.

The yeast population in *makai ko jaanr* and *gahoon ko jaanr* was found to be at the level of 10^7 cfu/g, whereas that of the LAB was below 10^5 cfu/g (Thapa 2001). Yeasts were dominant both in numbers as well as in their activities. Microorganisms associated with these native products were the same as those present in commonly used starter, such as *marcha*.

The naturally fermented alcoholic beverages of Himachal Pradesh contain mycelial fungi (*Aspergillus flavus, A. oryzae, Mucor* spp., *Rhizopus*

Table 8.3 Nutritional Composition of the Himalayan Alcoholic Beverages

Parameter	Kodo ko Jaanr or *Chyang*	Bhaati Jaanr	Gahoon ko Jaanr	Makai ko Jaanr
Moisture (%)	69.68	83.4	73.7	81.9
pH	4.08	3.5	3.9	3.3
Acidity (as % lactic acid)	0.27	0.24	0.37	0.38
Alcohol (%)	4.8	5.9	3.1	2.5
Ash (% DM) [a]	5.1	1.7	2.5	2.1
Fat (% DM) [a]	2.0	2.0	0.6	3.1
Protein (% DM) [a]	9.3	9.5	12.3	13.1
Crude fiber (% DM) [a]	4.7	1.5	10.5	2.8
Carbohydrate (% DM) [a]	83.7	86.9	84.7	81.8
Food value (kcal/100 g DM)	389.6	404.1	393.0	407.1
Minerals (mg/100 g DM)				
Calcium	281.0	12.8	18.3	5.2
Magnesium	118.0	50	102.0	70.0
Manganese	9.0	1.4	2.9	0.5
Copper	2.2	1.4	1.0	0.9
Iron	24.0	7.7	13.6	17.0
Zinc	1.2	2.7	1.6	1.2
Sodium	39.0	24.7	26.7	21.5
Potassium	398.0	146	300.0	227.0
Phosphorus	326.0	595	763.0	538.0

Note: Data represent mean values.

[a] % DM = g/100 g, on a dry matter basis.

spp.), yeasts (*Saccharomyces* spp., *Torula* spp., *Rhodotorula* spp.), and LAB (*Pediococcus, Leuconostoc, Pseudomonas,* etc.) (Joshi and Sandhu 2000).

8.5 Nutritive value

Marcha is slightly acidic in nature, with pH 5.6 and 0.1% acidity. Sun-dried *marcha* contains 14% moisture after fermentation and 1.4% ash (Tamang 1992). The acidic nature of *marcha* is due to the presence of high populations of LAB (Thapa and Tamang 2004). Given the low moisture content and acidic nature of *marcha*, it has a long shelflife and can be stored at room temperature for a year or more.

The nutritional composition of the Himalayan ethnic fermented beverages is given in Table 8.3. Due to cooking prior to fermentation, the moisture content is slightly higher in the fermented product. *Kodo ko jaanr* is a mildly alcoholic, sweet-flavored beverage. No remarkable changes were observed in the fat and protein contents of *bhaati jaanr* over the substrate. Crude fiber content, however, did increase during fermentation.

The Himalayan fermented cereal beverages are a high-calorie food, having more than 400 kcal (100 g/dry matter) of energy (Thapa and Tamang 2004), which is considerable for maintenance of bodily functions (Wardlaw et al. 1994; Basappa 2002). Because of their high-calorie content, these foods are favored by ailing persons and postnatal women, who consume the extract of *kodo ko jaanr* as well as *bhaati jaanr* to regain strength. Finger millet, called *ragi* in South India, is a good source of iron, calcium, magnesium, and phosphorus (Samantray et al. 1989). Ash, fat, and protein content remain the same as that of the substrate. Fermentation of finger millet enhances bioenrichment of minerals (Thapa and Tamang 2004). Fermentation of finger millet to *kodo ko jaanr* enhances bioenrichment of minerals such as Ca, Mg, Mn, Fe, K, P, contributing to mineral intake in the daily diet of the Himalayan rural people. An increase in mineral contents—mostly calcium, iron, sodium, potassium, and phosphorus—is also observed in *bhaati jaanr* due to fermentation (Tamang and Thapa 2006).

Chyang contains threefold the amount of B-vitamins (riboflavin, niacin, and pantothenic acid) and comparatively more folic acid than the nonfermented product. It is noteworthy that cyanocobalamin, which is not present in finger millet, is synthesized by the fermenting microorganisms (Basappa 2002). The essential amino acids—valine, threonine, leucine, and isoleucine—are in higher concentration in *chyang* (Basapa et al. 1997). The contribution of *chyang* (per liter) is very significant, as it fulfills 22%–2000% of the RDA per day of 1.5 mg thiamine, 1.7 mg riboflavin, 6.0 mg pantothenic acid, 18 mg niacin, 180 mg folic acid, and 20 mg cyanocobalamin of the B vitamins (Basappa 2002). The beverage prepared by starter cultures contained more thiamine, pantothenic acid, and niacin than that of *phab* (Basappa et al. 1997).

Table 8.4 Alcohol Content (% [v/v]) of *Raksi*, Ethnic Alcoholic Drink of the Himalayas

Distilled from *Gahoon ko Jaanr*	Distilled from *Bhaati Jaanr* Mixed with Rhododendron Flowers	Distilled from *Bhaati Jaanr*	Distilled from *Makai ko Jaanr*	Distilled from *Kodo ko Jaanr*
22.6	24.0	22.8	22.7	22.5
(22.5–23.3)	(23.1–26.5)	(22.7–23.1)	(22.5–22.8)	(22.5–22.5)

Note: Data represent the means of five samples from each source. Ranges are given in parentheses.

In *poko* of Nepal, a fermented rice beverage, thiamine was found to increase up to 16%–32%, pyridoxine by 50%–59%, vitamin B$_{12}$ by 18%–53%, folic acid by 76%, and niacin content by 117%–173% (Shrestha and Rati 2003). Carbohydrate decreased from 86.4% (in control sterilized rice) to about 77.3% in *poko* . The calorie value of *poko* is estimated to be around 4 kcal/g. The nutritive value of *poko*, including vitamin content, increases during the traditional fermentation process (Shrestha and Rati 2003; Dahal et al. 2005). *Zutho* contains 5% (v/v) alcohol, with pH 3.6 and acidity 5.1% (Teramoto et al. 2002).

The pH, acidity, and alcohol content of *raksi* are 3.6, 0.06%, and 22.9%, respectively (Thapa 2001). *Raksi* distilled from *bhaati jaanr* mixed with a few petals of *Rhododendron* shows the highest alcohol content (27% [v/v]), comparable with *raksi* prepared from other fermented cereals (Table 8.4).

8.6   Conclusion

Ethnic fermented beverages constitute an integral part of the dietary culture of the Himalayan people. The traditional method of subculturing of mixed inocula using rice as base substrates for preparation of starter is a remarkably innovative technology developed by the *marcha* makers using their native skills, which is a technique of subculturing an essential consortium of microorganisms. There is no authentic record of how long this subculturing technique has been practiced by the ethnic people, but it is certain that they have maintained the microbial diversity associated with the food ecosystem of the Himalayas for centuries, probably for the last 2500 years, as the ancient history of Nepal records the consumption of alcoholic beverages during the Kirat dynasty around 600 b.c.

The consortium of microorganisms in ethnic Himalayan alcohol-producing starters contains filamentous molds, enzyme-producing and alcohol-producing yeasts, and a few pediococci and lactobacilli. The nutritional composition of the Himalayan ethnic alcoholic beverages provides the required calorie content for a majority of the people.

An attempt was made to upgrade the traditional processing of *kodo ko jaanr* using selected strains instead of conventional *marcha*.

The anatomy of the traditional distillation apparatus designed by the ancient Himalayan women for production of ethnic alcoholic drinks perfectly fits the chemistry of distillation of modern days. What a wonderful innovation of indigenous science linked with culture!

chapter nine

Antiquity and ethnic values

9.1 Antiquity

Each ethnic food has its history or antiquity. Indian food habits pre-dating 3000 b.c. have been well documented based on historical documents and archaeological evidence (Yegna Narayan Aiyar 1953; Prakash 1987; Prajapati and Nair 2003). Similarly, the history of Chinese foods and cuisines has been written and compiled by several historians since 4000 b.c. based on historical monuments and records (Lee 1984; Yoon 1993). However, tracing back to the history or origin of ethnic Himalayan fermented foods is not possible due to the lack of authentic historical monuments, literature, and archaeological evidence. Records of agricultural and pastoral systems—including cultivation of cereals (rice, wheat), vegetables, cattle rearing, consumption of milk and milk products, meat, etc.—during the Lichchhavi dynasty in Nepal around 100–880 a.d. are available (Khatri 1987). The Lichchhavi kings of Nepal used to organize grand feasts for the Hindu priests around 100–880 a.d. (Khatri 1987), indicating the importance of traditional foods in ancient Nepal. The ancient historical monuments of Nepal indicate that the Himalayan ethnic foods have been consumed in the region for more than 2500 years.

The Himalayan fermented foods and beverages have been consumed since time immemorial and are part and parcel of each ethnic community originated, settled, or migrated within the Himalayan regions. The native skills of food fermentation in the Himalayas have been passed from mothers to daughters and from fathers to sons through the traditional knowledge of the elders, which includes grandmothers/grandfathers, mothers/fathers, village elders, self-practice, family tradition, community knowledge, neighbors, etc. The antiquity of food culture is linked with the cultural and political history as well as the ethnicity of the region. In this book, I have tried to put forward the bases for the history of some common Himalayan fermented foods based on the myths and unrecorded beliefs of the ethnic people. The documentation of ethnic information on the history of Himalayan fermented foods and beverages may provide vital information about the history and food culture of the different ethnic peoples of the Himalayas. The history (antiquity) of some common ethnic foods of the Himalayas is discussed in this chapter.

9.1.1   Antiquity of kinema

The domestication of soybeans first emerged in the eastern half of North China around the eleventh century b.c. during the Chou dynasty, which is considered as the primary gene pool of the soybean (Hymowitz 1970). The migration of the soybean from the primary gene center to South China, Korea, Japan, and Southeast Asia probably took place during the expansion of the Chou dynasty (Hymowitz 1970). Soybeans were introduced in Japan, Southeast Asia, and south central Asia in the first century a.d., which is considered as the secondary gene center for soybeans (Hymowitz and Kaizum 1981). Dissemination of soybeans was based upon the establishment of sea and land trading routes, the Silk Road, the emigration of some tribes from China, and the rapid acceptance of the soybean as a staple food by other ethnic communities (Hymowitz and Kaizum 1981). Within the secondary gene center, there are regions in which the soybean has been extensively modified (as in Japan), while in other areas the soybean has not been modified at all or, at the most, very slightly from the original introductions, for example, in northern India. Hence, Japan is to be considered a very active microcenter and northern India a passive microcenter within the soybean secondary gene center, and central India (Madhya Pradesh) may be considered a recent or tertiary soybean gene center (Hymowitz and Kaizum 1981). It is suggested that soybeans grown in central India were introduced from the secondary gene pool center, i.e., Japan, south China, and Southeast Asia, and that the soybeans grown in the northern part of the Indo-Pakistan subcontinent came from central China (Hymowitz and Kaizum 1981).

Many anthropologists and historians claim that ethnic fermented soybean foods of Asia might have originated from *douchi* or *tau-shi*, one of the oldest ethnic fermented soybean foods of China, during the Han dynasty in southern China around 206 b.c. (Bo 1984; Zhang and Liu 2000). Yoshida (1993) believed that production technology and consumption of *douchi* expanded northward to northern China, westward to eastern Nepal, southward to Indonesia, and eastward to Japan during the Han dynasty. The Kirat race (to which the Limboo belong) is believed to have come to their present abode in Limbhuwan in Nepal between 600 and 100 b.c. (J. R. Subba 1999). The period from 625 b.c. to 100 a.d. has been described by many historians as the rule of the Kirat kings in Nepal (Adhikari and Ghimirey 2000). The kingdom of Limbhuwan (presently the districts of eastern Nepal—Therathum, Taplejung, Panchthar, Dhankuta, and Ilam) was established by the Limboo kings (J. R. Subba 1999). The word *kinema* was originated from the word *kinamba* of the Limboo language (*ki* = fermented, *namba* = flavor) (Tamang 2001b). It is widely believed that the Limboo started its production and consumption, though there is no historical document on the origin of *kinema*. Whether *kinema* appeared first and diversified to other similar products, or vice versa, is a big challenge to food anthropologists, historians, and food microbiologists.

The Limboo believe that the origin of the soybean is mentioned in one of their oral myths known as *mundhuns* (Subba and. Subba, pers. comm.). As per the myth, once upon a time there was a Limboo king, Lilim Hang. During his reign, the country was facing a severe famine and he was desperately looking for food resources and supply. In a dream, the *Yuma*, the god of Limboo, directed him to instruct his subjects to look toward the east and collect and eat whatever they could find in the agricultural fields. In the morning the king awoke and looked toward the east, where he saw soybean plants. Immediately, he ordered his subjects to collect the seeds and eat. They never knew the seeds were soybeans. They saved their lives, and the famine was overcome. According to the *mundhuns*, the origin of soybeans and the habit of cultivation of soybeans as a part of the agricultural system of Limboo came into existence. They coined the term *chembi* for soybean. The Limboo believe that soybeans and other edible items after harvesting should be offered to their god *Yuma* to extend their gratitude for saving them from famine. It is a myth, and no record of *mundhums* is available. However, according to Subba (2008), the Limboo oral mythology in the form of *mundhums* might have been written between 2500–100 b.c.

Since it is proved that soybeans were first domesticated in the eleventh century b.c. during the Chou dynasty in China, which is considered as the primary gene pool of the soybean (Hymowitz 1970), the fermentation of soybeans into various recipes might have originated only after the eleventh century b.c. *Natto* was introduced to Japan from China by Buddhist priests during the Nara period around 710–794 a.d. (Itoh et al. 1996; Kiuchi 2001). *Kinema* might have originated in east Nepal around 600 b.c. to 100 a.d. during the Kirat dynasty.

The unification of Nepal in the seventeenth century a.d. resulted in the migration and movement of people from one place to another, and thus the exchange of culture and food habits within modern Nepal. *Kinema* is not traditionally eaten by the Nepali Brahmins. Although the reason is not documented, it is believed that the Brahmin consider *kinema* as stale food or not fresh food. Consumption of fresh food was a social compulsion of the upper-caste Hindu. Another reason of not eating *kinema* may be its umami-type flavor, which may not be appealing to consumers. However, for many ethnic non-Brahmin communities in the Eastern Himalayas, *kinema* and similar products have been adopted as a highly accepted delicacy in their food habit. The Lepcha call it *satlyangser*; the Tibetans and Bhutia call it *bari*; the Khasi call it *tungrymbai*; the Meitei call it *hawaijar*; the Mizo call it *bekang*; the Sema Naga call it *aakhone*; and the Apatani call it *peruyyan*. The similar sticky whole bean with an alkaline nature is *natto* of Japan, *chungkokjang* of Korea, and *thua-nao* of Thailand.

The nature of the products showed similarities due to stickiness, non-saltiness, and fermentation by bacilli (purely bacteria); however, some steps in production and culinary practices vary from product to product. The plasmid of *Bacillus subtilis* (*natto*) strain resembles that of *Bacillus subtilis* isolated from *thua-nao* and *kinema* (Hara et al. 1995). Based on phylogenetic analysis by determining the 16S rRNA sequencing, similarity among the strains of *B. subtilis* isolated from common sticky fermented soybean foods of Asia has been observed (Tamang et al. 2002). This suggested that *B. subtilis* strains might have originated from the same stock. If a hypothetical line is drawn, it becomes a triangle starting from Far East Japan (*natto*), touching Korea (*chungkokjang*); and then extending to South China (*douche*), eastern Nepal, Darjeeling hills and Sikkim (*kinema*); and finally extending to northern Thailand (*thua-nau*). Nakao (1972) termed this triangle as the "triangle of *natto*" and included all fermented soybean products, including miso, soy sauce, and tempeh, and extended the triangle up to Indonesia. I propose this hypothetical triangle may be renamed as the "*Kinema-Natto-Thua nao* triangle" (KNT-triangle) instead of the "*natto*-triangle" proposed by Dr. Sasuke Nakao. Within the proposed triangle-bound countries, many familiar as well as less familiar fermented, sticky, nonsalty soybean foods are consumed by the different ethnic groups of people in Cambodia, Laos, Vietnam, Northeast India, Myanmar, and southern parts of China (Figure 9.1). Beyond this hypothetical KNT-triangle, there is no report of *kinema*-like products with sticky and ammonia-flavored fermented soybean foods. The proposed KNT triangle does not include salted, sticky, and nonbacilli-fermented soybean products such as tempeh, miso, *sufu*, soy sauce, etc.

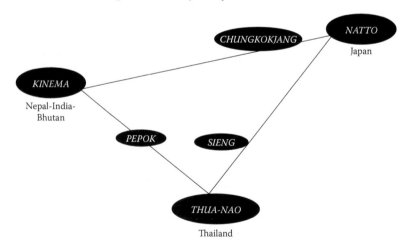

Figure 9.1 Kinema-Natto-Thua nao (KNT): Triangle of *Bacillus*-fermented sticky nonsalted soybean foods of Asia.

Another theory of the antiquity of fermented sticky soybean could also be that its discovery was accidental. Cooked soybeans could have been left after the meal, and the next morning viscous stringy threads with a typical flavor were noticed on the cooked beans. People might have tasted this product with appealing flavor and texture, and started liking the product. This could be the origin of *kinema* and other similar fermented sticky soybean foods, subsequently developed by further innovation and demand of local people, depending on agroclimatic conditions. The history of *kinema*, however, is yet to be fully studied.

It has been observed that mildly alkaline-flavored sticky fermented soybean foods are popular among the Mongolian-origin races. This may be due to its typical flavor, called umami flavor (Kawamura and Kara 1987), developed during proteolysis of soy protein to amino acids in fermentation. The Mongolian people like to have umami-flavored foods due to specific sensory development. This is the reason why fermented soybeans are prepared and consumed exclusively by the Mongolian-origin races in the Himalayas. However, it also depends on the cultivation of soybeans as a tradition, which is not common among the highland Tibetans due to the extremely cold climate. Hence, consumption of soybeans both fermented and nonfermented is uncommon among the Tibetans.

Alkaline sticky fermented soybean food is traditionally eaten directly without frying or cooking, since soybean is cooked prior to fermentation. This culture of eating freshly prepared fermented soybean is still common among the Japanese (*natto*), the Koreans (*chungkokjang*), the Sema Naga (*aakhone*), and other North Eastern Indians except *kinema* by the Nepali. *Thua-nao* is made into soup by boiling in Thailand. *Chungkokjang* is also consumed as a soup by the Koreans. *Kinema* is fried in oil and made into a curry. There is no record of eating *kinema* without cooking or frying. Probably the influence of Hindu food habits, by frying the foods in edible oil to make curry and also to make it more of a delicacy, has upgraded the culinary skills of *kinema* consumers. The culinary practices of preparing *kinema* is a blend of Aryan and Mongolian food culture. *Kinema* is one of the oldest cultural foods in Asia.

Though the Eastern Himalayan people have culturally adopted eating soybeans in both nonfermented and fermented products, traditionally they never prepare and consume miso, soy sauce, tofu, *sufu,* tempeh, etc. Moreover, they never prefer the soy milk due to its beany flavor and also due to the cultural acceptability of animal milk, particularly cow milk. Dry soybean seeds are roasted and eaten directly or made into pickles that are very common among the Nepali. Fresh soybeans are boiled and eaten. Probably the people did not bother to adopt the production method of miso, soy sauce, tempeh, and *sufu* due to handling of starter cultures separately and also the more complex steps in their production methods, in contrast to the *kinema*-making process, which is simple, labor-intensive, and naturally fermented. The other reason for not adopting non-*kinema* soybean products may be that there were several other foods in the traditional cuisine of the Himalayan people—selecting from milk to meat, vegetable to bamboo, etc.—and also the influence of the Hindu ethos and culture. Personal choice of foods and the time taken for production and the availability of raw materials might have compelled the people to limit their choice to *kinema* or sticky fermented soybean foods rather than miso- and soy sauce–type soybean foods.

9.1.2   Antiquity of gundruk and sinki

There is a myth about the invention of *gundruk* and *sinki* in the Nepali culture. During a time of war, attacks from enemies forced the farmers to flee the villages during the winter, when leafy vegetables, radish, and paddy were in plenty in the fields. As per war tactics, before the people fled, all food items, water holes, roads, bridges, and other facilities and amenities were destroyed as much as possible, so that advancing enemy troops could be crippled. However, the wise king had gone a step ahead in depriving the enemy while also preserving food stocks for his soldiers and farmers upon their return. The king ordered his subjects to cut ripening paddy and to uproot radish crops in the field, and then he directed them to dig pits and make beds of hay, to bury the prematurely harvested crops in separate pits, and then to cover them with hay and mud. The farmers and soldiers dug the pits and buried all agricultural process, including the radish, leafy vegetables, and paddy—whatever they could hide from the attackers. The king and his subjects returned to their homes after a few months and dug out the pits for paddy. However, the rice obtained from the paddy was stinky, while the radishes and leafy vegetables had developed an acidic and sour taste that was different from the fresh vegetables. After removal from the pits, a large amount of the fermented vegetables was kept lying on the field in open air. After a few days, the freshly fermented vegetables were dried and fully preserved, with a decrease in weight due to drying. The people liked the taste and flavor and termed these products as *gundruk* for leafy vegetables and *sinki* for radish. This is how *gundruk* and *sinki* have developed as preserved foods. They made the products into soup and pickles using their culinary skills. The people might have standardized the production method in their own way and invented the biopreservation technology of perishable vegetables. Primarily, people wanted to preserve the rice grains to stock for consumption and to keep away from the enemies, but the natural process had made the rice into an unacceptable product with a highly stinky flavor that was not appealing to the consumers. This they termed *hakuwa*, meaning the "bad spell," which could not be used as edible rice. On the other hand, they accidentally discovered the new fermented vegetable products *gundruk* and *sinki* with highly acceptable flavor and acid taste and started adopting this product in their food habits.

The word *gundruk* might have originated from the word *gunnu*, meaning "dried taro stalk" in the Newar dialect. Although there is no historical record, the invention of *gundruk* fermentation might have originated during the reign of the Newar in ancient Nepal. This is a myth; I researched the veracity of this myth, but could not get any documented literature, except an old Nepali vernacular rapid reader textbook on *sinki* and *hakuwa*, written by an eminent Nepali writer, the late Paras Mani Pradhan. Elder people, including my grandmother and mother, continue to retell this myth that recounts how the Nepali forefathers discovered the indigenous technology of biopreservation from perishable vegetables. However, this is probably an example of "accidental fermentation" that was discovered during the search for other edible foods by the ancient people.

The knowledge of the art of pickling vegetables or fermentation of vegetables is lost in antiquity (Battcock and Azam-Ali 1998). It may have been developed in Asia, as suggested by Pederson (1979), or in the Mediterranean region (Hulse 2004), but until more evidence is available, its origin will remain obscure (Steinkraus 1996). The Chinese laborers working on the Great Wall were eating acid-fermented mixed vegetables in the third century b.c. (Pederson 1979). Centuries ago, the Koreans developed *kimchi*, made from acid-fermented Chinese cabbage, radish, and other ingredients (Lee 1994). Similarly, in the Western world, cabbage was fermented to sauerkraut and cucumbers to pickles (Pederson and Albury 1969). In any event, this method of food preservation has been used for many centuries and is one of the important methods of food preservation still in use for vegetables and fruits where production by canning, drying, or freezing is not the method of choice (Vaughn 1985). Modern canned or frozen foods are too expensive or not easily available for most of the people living in underdeveloped and developing countries, and acid fermentation combined with salting remains one of the most practical methods of preserving and often enhancing the organoleptic and nutritional quality of fresh vegetables (Steinkraus 1996).

9.1.3 Antiquity of fermented milk products

Some of the historical records depict the development of a dairy system in ancient India. The mention of cow and the importance of milk products was referred to in *Rig Veda*, the oldest sacred book of the Hindus (Prajapati and Nair 2003). The use of *dahi* has been mentioned in ancient Indian scriptures such as Vedas, Upanishads, etc. (Yegna Narayan Aiyar 1953). It is well known in ancient Indian history that *dahi* (buttermilk) and *ghee* (butter) were widely consumed milk products during Lord Krishna's time, about 3000 b.c. (Prajapati and Nair 2003). Cattle (cow) rearing was one of the important pastoral systems during the Gopala dynasty in Nepal, one of the earliest kingdoms in ancient Nepal in 900–700 b.c. (Adhikari and Ghimirey 2000). However, the history of ancient Nepal records that the main pastoral system during the Mahishapala dynasty in Nepal was buffalo rearing instead of cow in 700–625 b.c. (Bista 1967). Hence in modern Nepal, both cow and buffalo rearing are common practices among the farmers. The Aryan-Hindu pastoral system has influenced the preparation and consumption of milk and milk products in the early settlement in the Himalayas. Nepal was a Hindu kingdom until it became republic in 2007 and has a deep-rooted Hindu culture, where the cow is regarded as a sacred animal and its milk and milk products are used in every religious and cultural function. Many ancient Indian histories mention the origin of *dahi*—one of the oldest fermented milk products of the Hindus—and other fermented milk products during the period from 6000 to 4000 b.c. (Yegna Narayan Aiyar 1953). Milk is not traditionally consumed in China. Hence the milk and milk products of the Himalayas might have originated from the main Indian Hindu culture. The Tibetans living at high altitudes have long been using yak milk and its different fermented products (*chhurpi, chhu, philu*, etc.), unlike the Chinese, who do not traditionally consume milk and milk products.

9.1.4   Antiquity of selroti

Selroti is an important fermented cereal-based food in the local diet of Nepali of the Himalayas. *Roti* is a Nepali word for bread. Out of many kinds of bread the Nepali consume, two of them have a special place in the society. One is *babari*, and the other is *selroti*. *Babari* is a round, solid pancake, whereas *selroti* is a ring-shaped doughnut-like bread. Both are prepared from pounded rice flour. It was recorded during an interview that, in olden days, only *babari* was prepared and consumed by the people instead of *selroti* (Yonzan 2007). As the story goes, the consumers found it difficult to fry in a pan, especially to turn it upside down; so they started making rings with the batter. And to turn ring-shaped bread, people started using a poker, locally called *suiro* (a pointed bamboo stick). Anything lifted with a *suiro* is called a *saela* in the Nepali language. The word *selroti* probably is derived from the word *saela*, meaning lifting any item. In this case, lifting could indicate the lifting of deep-fried *selroti* from a pan containing hot edible oil by a poker. That could explain how preparation and consumption of *babari* among the Nepali was slowly replaced by *selroti* preparation, which has become a distinct food culture of the Nepali.

There is also another hypothesis on the nomenclature of *selroti*. The word *seli* is a name for a local variety of rice cultivated in the foothills of Nepal. The product prepared from *seli* might have been called *selroti*, since *selroti* is prepared during *tiwar*, one of the main festivals of the Nepali that is celebrated once a year or *saal* (Nepali word meaning "a year"). Some people believe that the word *selroti* is derived from the word *saal*. The antiquity of *selroti* remains a myth, and no historical documents or monographs are available on this product.

Jalebi has been known in northern Indian areas since 1450 a.d. and is probably of Arabic or Persian origin (Gode 1943).

9.1.5 Antiquity of alcoholic beverages

Jaanr was mentioned in the history of Nepal during the Kirat dynasty (625 b.c.–100 a.d.) (Adhikari and Ghimirey 2000). The Newar used to ferment alcoholic beverages from rice during the Malla dynasty in 880 a.d. (Khatri 1986). Risley (1928) wrote *"marwa* or *chyang*, is a kind of beer brewed by everyone in Sikkim, and might be called their staple food and drink." In the old literature of Darjeeling and Sikkim written in 1854, there are brief descriptions of fermented millet beverages of Darjeeling hills and Sikkim (Hooker 1854; Gorer 1938). The fermented beverage is also known as *chyang* by the Bhutia (Risley 1928) and *chi* by the Lepcha (Gorer 1938). *Chyang* has been reported from the Ladakh region by Bhatia et al. (1977). *Thumba* (correctly spelling is *toongbaa*), the fermented beverage common in Darjeeling hills, Sikkim, and Nepal, has been reported by Hesseltine (1965, 1979) and Batra and Millner (1976). Actually, *toongbaa* is a vessel made of bamboo shoots to contain the grits of fermented finger millets *kodo ko jaanr* or *chyang*; however, the term *toongbaa* is popularly used to denote the fermented beverage, too.

Because ethnic fermented foods were produced by natural fermentation, the origin of cereal fermentation technology is obscure (Haard et al. 1999). In Asia, the malting process is rarely used in traditional fermentation processes. Instead, fermentation starters prepared from the growth of molds on raw or cooked cereals are more commonly used (Haard et al. 1999). The use of mixed starters might have its origins during the time of Euchok, the daughter of the legendary king of Woo of China, known as the Goddess of rice-wine in Chinese culture, in 4000 b.c. (Lee 1984). The first documentation of *chu*, the Chinese starter, similar to *marcha* of the Himalayas, was found in Shu-Ching documents written during the Chou dynasty (1121–256 b.c.), in which it is stated that *chu* is essential for making alcoholic beverages (Haard et al. 1999). According to Chi-Min-Yao-Shu documents written by Jia-Si-Xie of the late Wei kingdom in the sixth century b.c., many methods of preparation of *chu* were described (Yoon 1993). The use of *chu* for rice-wine production was commonly practiced in the spring and fall and Warrior Periods of China during the sixth to seventh centuries b.c. and the beginning of the Three Nations' Periods in Korea during the first century b.c. to the second century a.d. (Lee 1995). It might have transferred from Korea to Japan in the third century a.d. according to Kojiki, or Chin, whose memorial document is kept in a shrine at Matsuo Taisha, Kyoto, Japan (Lee 1995).

According to the historical document written by Chi-Min-Yao-Shu (530–550 a.d.), *chu* was prepared from barley, rice, and wheat (Yoon 1993). Ten different types of *chu* were described in the Chi-Min-Yao-Shu monument (Yoon 1993), all of which were used for the fermentation of alcoholic beverages in China. The cake-type *ping-chu* is identical to *nuruk* of Korea, and the granular-type *san-chu* is similar to Japanese *koji* (Yoon 1993). According to Yokotsuka (1985), *chu* may either be yellow (*huang*), possibly due to *Aspergillus oryzae*, or white, probably due to *Rhizopus* and *Mucor*. *Nu-chu* is prepared from cooked rice, which is shaped into a cake and then cultured with molds (Yokotsuka 1985). Wheat *chu* originated in North China and the Korean Peninsula, while rice *chu* originated in South China (Haard et al. 1999). The process of cereal alcohol fermentation using mold starters was well established by the year 1000 b.c., and 43 different types of cereal wines and beers were described with detailed processing procedures in Chi-Min-Yao-Shu (530–550 a.d.) (Yoon 1993, Haard et al. 1999).

Marcha or *phab* of the Himalayas might have originated from South China during migration of Mongoloid races to the Himalayas. This is based on the similarity of the *nu-chu* and *marcha* making processes and the fact that the filamentous molds consisting of Mucorales are the dominant mycelial fungi in *marcha*, whereas aspergilli are restricted only to *mana* prepared from wheat in Nepal. The exact year of origin of *marcha* making in Nepal and *phab* making in Tibet is unknown.

9.1.6 Antiquity of fish products

Historically, lactic fermentation of fish was associated with salt production, irrigated rice cultivation, and the seasonal behavior of fish stock (Lee et al. 1993). The Mekong basin was most probably the place of origin of these products, and the Han Chinese (200 b.c.–200 a.d.) learned of the process when they expanded south of the Yangtze River (Ishige 1993). Fish products prepared by lactic acid fermentation remain common in Laos, Cambodia, and in the north and the northeast of Thailand (Ishige 1993). Unlike Chinese-type fermented fish, the Himalayan fish products are slightly different and mostly dominated by drying and smoking processes. Fermentation of fish is restricted to *ngari* and *hentak* in Manipur and *tungtap* in Meghalaya; the rest of the fish products are dried or smoked. Fish is not a common item of food in the Himalayan cuisine, since fish dishes are mostly associated with the people living nearby rivers or lakes. Most of the lands in the Himalayas are highlands and hill terrains. Moreover, meat eating is more prevalent than the fish eating among the nonvegetarian people in the Himalayas.

9.2   Ethnic values of the Himalayan fermented foods

9.2.1   Cultural foods

Bhagavad Gita, the sacred book of Hindu, divides food into three categories that correspond to quality and property. These are (a) *Sattvika* food for longevity, intelligence, strength, health, and happiness, which includes fruits, vegetables, grains, cereals, sweets, etc.; (b) *Raajasika* food for activity, passion, and restlessness, which includes hot, sour, spicy, and salty foods; and (c) *Taamasika* food, which includes intoxicants and unhealthy foods that cause dullness and inertia. We grade the Himalayan fermented foods as per the Hindu philosophy: Almost all fermented foods of the Himalayas are *Sattvika* foods due to their high nutritive value, functionality, and health-promoting benefits; some acidic foods, such as fermented vegetable and bamboo products, are *Raajasika* foods due to their sour taste; none of the Himalayan foods have a negative effect on health and mind, and hence these are not *Taamasika* foods.

Ethnic fermented foods and beverages in the Himalayas are socially and culturally acceptable food items in the local cuisines. Besides taste and aroma, ethnic foods have been used for nonedible purposes such as worshipping gods and goddesses, offering to nature, spirit possession, etc. Some ethnic foods have social importance for celebration, particularly during festivals and special occasions. Every food has its social or ethnic value and is associated with the customs and culture of the community. There exists a ritual that the Nepali Brahmin male after *bratabandha* (formalized into Hindu) and the Hindu female after marriage are not allowed to take stale food (Tamang 2009). As per the conventions, food cooked in a roofless place is not allowed to be eaten, and this belief still prevails in the society. The Himalayan kingdoms like Sikkim and Nepal were ruled by kings for a period of time; as a result, there emerged only two classes of people in the society: the rich and the poor (Tamang 2007b). Ethnic fermented foods like *kinema*, *gundruk*, and *sinki*, along with boiled maize pudding, locally called *dheroh*, are generally graded as poor-man's foods and are common among the rural poor (Tamang 1997); the rich people during that period might not have tasted or consumed such foods due to social status. Probably, these ethnic foods must have had their origin in the food palate of the poorer section. The poorer class included the working laborers and farmers, who needed inexpensive and easily cooked food to fill their tummies to the fullest. Thus, foods like *dheroh*, *kinema*, *gundruk*, *sinki*, etc., might have emerged as a result of this need and preference of the poor class. These foods are still considered as poor-man's foods in the Nepali society. Today, these ethnic foods are associated with culture and tradition, yet they are not at all accepted to be consumed during special occasions and festivals.

Bhutan is the only living Himalayan kingdom. The Bhutanese people have more or less homogenous food habits and are mostly a nonvegetarian populace. Their cuisine is similar to the Tibetan cuisine. However, the ethnic Nepali of Bhutan have identical food habits as the people of Darjeeling hills, Sikkim, and Nepal. The fusion of ethnic foods of the Nepali and the Drukpa has produced a unique Bhutanese cuisine.

9.2.2   Fermented vegetables

The invention of biopreservation methods by the ethnic people of the Himalayan regions through pit fermentation or lactic acid fermentation is significant due to the ability to transform the available raw materials at a season of plenty to food materials during a season of deficit. Moreover, this is a remarkable step to store perishable vegetables in the absence of cold storage or refrigeration, where the majority of rural people cannot afford modern canned or frozen foods. Sun drying of freshly prepared *gundruk* and *sinki* is a traditional preserving method by which the shelf life of the products is prolonged. Dried products are preserved for several months without refrigeration and consumed during the long monsoon season, when fresh vegetables are scarce. Dry *gundruk*, *sinki*, and *inziang-sang* are comparatively lighter than the weight of fresh substrates and can, therefore, be carried easily while traveling a long distance in the difficult terrains of mountainous country like Nepal. People might have invented such preservation techniques to feed themselves while traveling for long distances, and also to store for future use whenever vegetables were in plenty. Carrying dry fermented products is still a common practice in the Himalayas while traveling for long distances (Tamang 2005b). Because of the acidic taste, *gundruk*, *sinki*, and *inziangsang* are said to be good appetizers, and the ethnic people use these foods for remedies from indigestion. *Gundruk* and *sinki* are part of the ritual codes that apply after a death in the family for the Newar. They believe that *gundruk* and *sinki* may not be eaten for the first eleven days following a death in the family. On the last day, i.e., the 20th day of the death ceremony, *gundruk* and *sinki* are traditionally served to the family. This is a tradition that is still followed by the Newar. *Soibum* based dishes are highly preferred during social and religious ceremony in Manipur.

9.2.3   Fermented milks

Dahi (curd) plays an important part in the socioreligious habits of the Himalayan people and is considered as a sacred item in many of their festivals and religious ceremonies both by Hindu and Buddhists. *Dahi* is a common savory, but it is also used as an adhesive to make *tika* with rice and as a colored powder that is applied to foreheads by the family elders during the Hindu festival, called *Dashai*. *Dahi* is also mixed with beaten rice, locally called *chiura*, and makes an essential food item during festivals such as *Ashar ko Pandra*, signifying the beginning of work in the fields for the farmers. *Dahi* is offered to the bridegroom as a symbol of good luck during marriage and is an essence to solemnize Hindu marriages. The Tibetans, Bhutia, and Lepcha also use *dahi* (*shyow*) in their religious and social events in marriages and funerals. *Shyow* (curd) is served exclusively during the Tibetan festival of *Shoton*. *Mohi* or buttermilk is used as a beverage in meals during many social festivals and religious events. It is offered to guests and visitors in many of the homes as a refreshing savory. For both Hindus and Buddhists, *gheu* or butter is a sacred item in all their religious ceremonies, and it is used as an offering at times of birth, marriage, and death as well as during prayer. The Nepali have a tradition of giving *gheu* to newborns, along with honey given by the father, to protect them from disease. *Gheu* is also used for lighting the lamps for gods and goddesses in Hindu temples and Buddhist monasteries. Soft *chhurpi* is served as an important dish as a curry and as the pickle *achar* in many religious and social festivals. *Chhu* is an important local food and is consumed by the Tibetans and Bhutia as a soup along with rice when other foods are not easily available. Hard *chhurpi* is eaten in high altitudes by Tibetans as a chewing gum and masticator that gives extra energy to the body by continuous movement of the jaws as they chew the gum. *Somar* is consumed mostly by the older generation of the Sherpa and is generally consumed to increase the appetite and to cure digestive problems.

9.2.4 Fermented cereals

The celebration of festivals with *selroti* is a custom of the Nepali, and the food is served as a confectionery during festivals. *Selroti* is prepared in almost all functions and festivals, particularly on *BhaiTika in Tihar* by Nepali sisters for their brothers. It is also served in marriages and death-related rituals. It is customary to hand over a basketful of freshly fried *selroti* to a bride's parents by the groom during marriage among the Nepali. This is probably to supplement the sweet dish, which is traditionally not common among the Nepali, for greetings. Traditionally, a newly married Nepali bride visits her parents' house once in a year. When she returns back to her husband's house, she should carry a *thumsey* (local name for bamboo basket) that contains freshly fried *selroti*. This tradition is known as *pani roti* by the Nepali. *Selroti* is traditionally served along with other traditional food items during *Bhai tika*, a Hindu festival that is observed to honor the brothers by their sisters. Beside this, it is also served during other festivals of the Nepali like *chaitay dasai*, *Maghay Sakranti*, *bara dasai*, etc. Fried *selroti* products can be preserved for about 10–15 days without refrigeration and consumed as is or slightly warmed up. People might have invented such preservation techniques to feed themselves while traveling. Carrying fried *selroti* is a traditional practice in the Himalayas while traveling for long distances.

Some of the ethnic fermented products of Himachal Pradesh have ethnic values. *Siddu* is served hot with *ghee* (butter) or chutney (pickle) in rural areas of Himachal Pradesh as a special dish on customary occasions. *Chilra* is prepared during marriage ceremonies and festivals in Himachal Pradesh. *Marchu* is eaten during local festivals (*phagli, halda*) and religious and marriage ceremonies in Lahaul. It is customary for a daughter to take *marchu* whenever she visits her maternal home from in-laws or vice versa in Himachal Pradesh (Thakur et al. 2004).

9.2.5 Fish products

Fish are consumed in Himalayan communities located near rivers and their tributaries, lakes, and ponds. The ethnic people of North East India catch freshwater fish from the Brahmaputra River and its tributaries and lakes. Nonconsumption of fish products by the Tibetans may be due to a religious taboo, as fish are worshiped for their power to bestow longevity. Similarly, the Buddhists strongly believe that their longevity may be prolonged through prayers (*tshethar*) after releasing captured fish into the rivers. Moreover, lakes are regarded as sacred by the Buddhists in Sikkim, Bhutan, Ladakh, and Tibet, barring them from fishermen. Another reason for not consuming fish in the local diet may be due to a taste for animal meats and dairy products, and to the difficulty in removing the small bones and scales of fish, which make them more difficult to eat than animal meats with larger bones.

9.2.6 Fermented meat

A sizable proportion of the populace in the Himalayas are meat eaters; however, regular consumption of meat is expensive for a majority of the poor people. People slaughter domestic animals (goats, pigs, cow, yaks, and sheep), usually on special occasions—festivals and marriages. During *dasai*, goats are ritually sacrificed to please the goddess Durga by the non-Brahmin Nepali. After the ceremony, the fresh meat is cooked and eaten as a family feast; the remaining meat is preserved by smoking above an earthen oven to make *suka ko masu* for future consumption. Tibetans, Bhutia, Drukpa, and Lepcha slaughter yaks occasionally and consume the fresh meat; the remaining flesh of the meat is smoked or preserved in open air, and this food is called *satchu*. The ethnic people of the Kumaun Himalayas prepare *chartayshya* curry, especially during the *kolatch* festival (worshiping the ancestral spirit), and offer this food to ancestors before eating (Rai et al. 2009).

9.2.7 Marcha

Marcha makers are restricted to the Limboo and Rai castes of the Nepali, and the Lepcha. Traditionally, the Limboo and Rai are known as *matwali*, meaning alcohol drinkers. The preparation of *marcha* is done by women of these castes. In order to keep this art secret, *marcha* is usually prepared at night. The trade in *marcha* is protected as a hereditary right of these castes. This may be the reason of adopting the *marcha* preparation only by certain ethnic groups. Marital status is a strong determinant in the preparation of *marcha* by the Rai castes of the Nepali, who allow only widows or spinsters to make *marcha*, whereas the Limboo do not follow the marital status for making *marcha*. Many *marcha* producers believe that addition of wild herbs gives more sweetness to the product, and they also believe that adding chillies and ginger get rid of devils that may spoil the product during preparation. This is actually to check the growth of undesirable microorganisms that may inhibit growth of native microorganisms of *marcha* (Soedarsono 1972), and addition of sweet herbs serves to supplement the carbon source for growing organisms in *marcha*. Moa tribes believe that the use of charcoal pieces and *Elscholtzia blanda* acts as an antimicrobial regulator to protect fermenting rice from contamination during *khekhrii* preparation in Nagaland (Mao 1998).

9.2.8 Alcoholic drinks

Drinking of alcohol is a part of the social provision for a majority of the ethnic people of the Himalayas, with the exception of Brahmin Hindu and Muslims, for whom alcohol is taboo. *Jaanr* and *raksi* are essential to solemnize the marriage ceremony of non-Brahmin Hindu Nepali and the Buddhist tribes. Eloping is a common practice in the Himalayas. Traditionally, after 3 days, relatives of the boy visit the girl's parents with bottles of *raksi/aarak* (a locally prepared ethnic distilled liquor) to respect the verdict of her parents and to pay the penalty for elopement. Once the consent is granted by the girl's parents, freshly prepared *raksi* is served to signify the union of two families, and the marriage is thus solemnized. Such practice of bridging between two families by a bottle of alcoholic drink is common only among the Himalayan people, mostly the non-Brahmin Nepali.

Ethnic alcoholic beverages have strong ritual importance. Alcoholic beverages are offered to perform the *pitri puja* or *kul puja*, the religious practice of praying to family gods and goddesses by the different ethnic people. *Mong chee* (alcoholic beverage of the Lepcha) is essential in various cultural functions of the Lepcha, such as *lirum*, *sejum*, and *namsung*. *Mandokpenaa thea* (alcoholic beverage of the Limboo), filled in *toongbaa*, and rice-made *raksi* are used for performing a ritual called *tonsin mundhum*. Those who come to offer condolences gathered at a funeral or a memorial service for the deceased are served with alcoholic beverages, mostly among Tibetans, Drukpa, Sherpa, Bhutia, Lepcha, Naga, Bodo, etc. Spirit possession among the ethnic people of the Himalayas is a common tradition (Hitchcock and Jones 1994). *Phedangma* and *bijuwa*, the Limboo priests, and *Lama*, the Buddhist priest, essentially use freshly distilled liquor *raksi/aarak* during spirit possession.

Losar (the Tibetan New Year) is celebrated by the traditional Tibetan cuisine, which basically consists of *chyang* (fermented barley beverage), fresh roasted barley flour for *phye-mar* (sweetened barley flour, symbolizing good wishes), *gro-ma* (a small, dried, sweet potato), *bras-sil* (sweet rice), and *lo-phud* (a young sprout of wheat or barley, symbolizing the birth of the new year), along with tea, butter, sheep's head, butter lamps, fruits and sweets, and especially locally produced foodstuffs (Rigzin 1993). *Ayela*, a clear drink of the Newar, is traditionally prepared during the month of December, immediately after the marriage season is over, which usually falls in November. The Newar believe that the quality of *ayela* is superior if it is prepared by a woman. The month (called *pahela* by the Newar, the month of fermentation, which is regarded as the best month for fermentation) is closely associated with *ayela*. This month falls in the winter season, shortly before the marriage season (December to mid-January). There is a superstition among the Newar that the quality of *ayela* determines the reputation of the marriage ceremony. If the best quality *ayela* is served, the guests may grade the ceremony as the best.

9.3   Conclusion

Many Himalayan ethnic fermented foods and beverages are similar to the Chinese ancient fermented foods and beverages. Examples include alcoholic beverages, mixed starter culture, sticky bacilli-fermented soybean, and acid-fermented vegetables. The Himalayan ethnic fermented foods and beverages that are dissimilar to the Chinese ancient fermented foods include fermented milk products, fermented meat products, fermented or dried/smoked fish products, fermented cucumber, and fermented cereal products.

The influence of the Chinese cuisine on the Himalayan food culture is mainly due to its common boundaries with Nepal, Bhutan, and India (Sikkim, Arunachal Pradesh, Uttarakhand, Himachal Pradesh, Jammu & Kashmir) as well as the common stock of races and ethnicity, i.e., the Mongolian origins. However, the food culture of the Himalayas is unique and is a fusion of the soybean/alcohol-consuming Chinese culture from the north and the milk/vegetable-eating Hindu culture from the south. In fact, the Muslim food culture has not much infiltrated into the Himalayan food culture except in the Jammu & Kashmir regions.

The British have influenced the food culture of the Himalayan people, mostly in the regions where the British ruled for a century or more, particularly in Darjeeling hills, Himachal Pradesh, Uttarakhand, Meghalaya, Assam, etc. However, the influence of British cuisine has not changed or added any new fermented food products to the Himalayan people. On the other hand, there are many English food products that have been adapted to the food culture of India, such as cakes, pastries, breads and loaves, processed cheese, hams, fruit juice, wine, beer, brandy, whiskey, rum, etc.

The Himalayan fermented foods cover all types of available substrates, ranging from milk to alcohol, legume/soybeans to cereals, vegetables to bamboo, meat to fish, etc. Although the history shows that some of these Himalayan fermented foods and beverages might have origins in China, the Himalayan people have altered the technology of preparation as per their choice and need. They have changed the cuisines for more delicacy, taking recipes from other races or communities, since they exist in a mixed-culture and heterogeneous society, but the end result is a distinct food identify of the Himalayas that is ethnic in nature.

The historical record of consumption of milk and milk products in Nepal in 900 b.c. (Adhikari and Ghimirey 2000) throws light on the cultural history of the food habits of the Nepali and other Himalayan people. The use of hands for feeding during the Lichchhavi dynasty from 100 a.d. to 880 a.d. (Bajracharya and Shrestha 1973) and historical monuments mentioning the grand feasts organized by the Lichchhavi kings from 100 a.d. to 880 a.d. (Khatri 1987) are some of the authentic historical records of the food culture of the Nepali. The diversity of ethnic fermented foods and beverages is more prominent in the Eastern Himalayas than that of the Western Himalayas. Among the ethnic preferences, the Nepali are the largest stakeholders as well as consumers of 80% of the Himalayan fermented foods and beverages. This is mainly due to the ethnic diversity within the Nepali community—from Brahmin to Kirat, from Aryan to Mongoloid—and to the projection of indispensable food culture from the single community of the Nepali (Gorkha).

The diversity of the ethnic Himalayan foods may be unknown to most of the countries outside the Himalayas, but the historical records show that these ethnic foods and alcoholic drinks have been consumed by the Himalayan people for more than 2500 years. These Himalayan fermented foods and beverages are indeed heritage foods.

chapter ten

Prospects of the Himalayan fermented foods

The Himalayan fermented foods have biological importance because of their ability to harness the resources of functional microorganisms, resulting in enrichment of bionutrients, biopreservative properties, medicinal and therapeutic properties, probiotic properties, etc. (Tamang 2007a). These ethnic fermented foods products provide a food safety net for the people during times of famine or environmental or human-made disasters. These foods are also a source of revenue for some mountain people to sustain their livelihood.

The ethnic fermented foods of the Himalayas can be used to enhance the regional economy and to promote sustainable development in the biodiversity-rich Himalayan regions. One can compromise on the quality of any product except food, which will have a direct impact on the consumer's health. Some ethnic Himalayan foods are more popular than others. These popular Himalayan foods could be introduced to consumers in other parts of the world.

Consumers prefer foods that are tasty and flavorsome, safe and healthy, with health-promoting and even remedial benefits, so that they might avoid taking extra oral medicines. Some of the Himalayan ethnic fermented foods have beneficial properties that can be a supplement for medicine and also serve to add taste and impart aroma. The possible prospects for commercialization of Himalayan fermented foods are based mainly on the following important parameters:

1. Medicinal aspects
2. Total substrate utilization
3. Food tourism in the Himalayas
4. Microbial genetic resources

10.1   Medicinal aspects

The food habits of the Himalayas typically exclude additional medicines or supplementary drugs. Most of the ethnic foods, both fermented and nonfermented, have therapeutic values and are eaten for prevention of illness. Traditionally, the Himalayan ethnic people do not have the habit of taking drugs and medicines in the form of tablets, tonics, etc. This may be due to the therapeutic values of their low-cholesterol ethnic foods, which contain antioxidants, antimicrobials, probiotics, essential amino acids, bionutrients, and some important bioactive or other beneficial compounds (Tamang 2007a).

The ethnic people who consume these foods believe that they have medicinal properties. *Somar* is used to cure stomach upset and to control diarrhea by the Sherpa. *Gundruk* and *sinki* are used to treat indigestion and are commonly eaten as an appetizer. The Meitei believes that *soibum* has wound-healing and antitumor properties for animals, particularly cows. *Kinema* is highly nutritive and is eaten to boost protein intake and to cure heart diseases. *Jaanr* and ethnic fermented beverages are high-calorie food beverages, rich in mineral content and vitamins, mostly given to post-natal women. *Bhaati jaanr* and *poko* promote good health, nourishing the body and giving good vigor and stamina. Ethnic Himalayan milk products have protective and probiotic properties that stimulate the immune systems and cure stomach-related diseases (Dewana 2002; Dewan and Tamang 2006, 2007). *Raksi* is a stimulating alcoholic drink that has both social and medicinal importance in the food culture of the Himalayan people. Although clinical study of the Himalayan fermented foods is yet to be carried out, the people who consume these customary foods believe that they have therapeutic value, and these products have been used both as foods and therapy for centuries. Such ethnic foods, if studied properly, may be find a place in the global marketplace.

10.2   Total substrate utilization

Unlike boiled rice and maize or wheat, finger millet or *kodo* is not so popular among the consumers in the Himalayas. Agriculturally, cultivation of finger millet is cost effective and nonlabor intensive compared with other crops like paddy, which need rapid and constant attention and proper irrigation. Finger millets, once sown in the hill terrain, do not need much care and can adapt to the agroclimatic conditions of the Himalayas. Nutritionally, finger millet seeds have a higher mineral content than other cereals (Gopalan et al. 2004). With such high calorie and mineral contents, low-cost millets have been converted into high-calorie, nutritious, flavorsome, palatable, and mildly alcoholic food beverages by the elders who invented the food fermentation technology using the mixed starter culture. The products are named *kodo ko jaanr* or *chyang* or *chee*. Once the extract has been sipped away, the remaining grits are used as animal fodder, given to cattle, pigs, and yaks, thereby contributing to the total utilization and biodegradability of the substrate. The fodder is also rich in bionutrients for the health of domesticated animals. This is a good example of total substrate utilization of low-cost cereal as a human food as well as an animal fodder through indigenous food fermentation technology in the Himalayas.

10.3   Commercialization through ethnic food tourism

Tourists visit new places and countries, and tourism is becoming increasingly popular all over the world. Tourist visits to a few hill resorts in the Himalayas has been increasing year by year. The concept of "ethnic food tourism" may have relevance in the present day due to the growth of the tourist industry in the Himalayas. The interaction of tourists with the people of other regions/countries, exploring their traditional values and culture, will extend to the enjoyment of dining on the local cuisine. Ethnic foods harness the cultural history of a community. Tourism provides an opportunity to demonstrate the value of indigenous knowledge in food production, highlighting its nutritious qualities, functional microorganisms, regional economy, and enjoyment of local cuisines.

The major components of a food-tourism marketing campaign in the Himalayas should focus on: the food tourism product as an attraction in destination marketing, consumer behavior, ethnic food production sites, and local markets where ethnic foods are sold (Hall et al. 2003). France attracts the highest number of tourists worldwide, reaching 82 million, more than its population of 64.5 million (UNWTO 2008). The secret is the delicious food and wine of France, served in inexpensive traditional restaurants that offer the local agricultural produce to millions of tourists (Kimura 2000). The main cuisine of France is cheese and wine, both fermented products. There are more than 5000 restaurants in Paris serving traditional dishes along with French wine. Tourists can also visit the many vineyards in France and learn how wines are made while sampling a number of different types and years of wines that were made at the vineyard—a trip that revolves around wine tourism. Tourists who travel to enjoy the foods of a region are the vanguard of a food culture that seeks to sample the cuisine of other regions. Tour promoters should focus on the specific food culture of a region in a form that appeals to such tourists.

The Himalayan ethnic foods can be diversified into more presentable forms to attract tourists to taste the aroma, flavor, and texture of unique recipes. Promotion of food tourism would require the opening of more ethnic food restaurants, food huts, and stalls, all focusing on traditional utensils, kitchen wares, and traditional culinary practices and customs while advocating the importance of the ethnic fermented foods. Tourists could be encouraged to visit villages, where they would learn of the traditional processing of Himalayan foods, similar to tourists who visit the vineyards of France and the production facilities for tempeh in Indonesia and *shoyu* and *sake* in Japan. Possible tourist sites for experiencing traditional foods in the Himalayas are: *kinema* production at Aho village in Sikkim; *kargyong* and hard-variety *chhurpi* making in North Sikkim; pit fermentation of *sinki* production at Kalimpong; *marcha* making in Therathum village in Nepal; *chyang* production in Ladakh; *ngari* production in Manipur; *mohi* production in Nepal, etc.

Village tourism has been increasing in the Himalayas. Serving a standardized version of ethnic Himalayan foods and drinks to tourists and travelers in village resorts would not only boost the regional economy, but also enhance local agricultural and livestock production due to the increased demand. Marginal farmers and local sellers would benefit. The model of food tourism in urban areas can be translated to local villages. Such tourists are eager to learn more about the culinary skills of local peoples, and their reports could promote the popularity of Himalayan foods outside the local regions.

10.4   Microbial genetic resources

Importance of microbial diversity as a genetic resource for application in food production, medicine, agriculture, and environment management has increased in recent years. Food bioresource is a basic part of an ecosystem where nature has given humans the ability to select foods from the available natural resources which are edible and culturally acceptable. Ethnic fermented foods and beverages are some of the important sources of microbial diversity in the Himalayas. Most of the ethnic fermented foods and beverages collected from different places in the Himalayas have been extensively studied and the functional microorganisms associated with ethnic foods isolated, characterized, and identified (Batra and Millner 1976; Chettri and Tamang 2008; Dewan and Tamang 2006, 2007; Jeyaram et al. 2008a, b; Karki et al. 1983d; Nikkuni et al. 1996; Rai 2008, Sarkar et al. 1994, 2002; Shrestha et al. 2002; Tamang et al. 2000, 2005, 2007a, 2008; Tamang and Nikkuni 1996; Tamang and Thapa 2006; Tamang and Sarkar 1993, 1995, 1996; Tamang and Tamang 2009b; Thapa et al. 2004, 2006; Thapa and Tamang 2006; Tsuyoshi et al. 2005; Yonzan 2007). The identified functional microorganisms isolated from the Himalayan fermented foods and beverages are listed below:

Lactic acid bacteria: *Enterococcus faecium, E. faecalis, E. durans; Lactobacillus alimentarius, Lb. amylophilus, Lb. bifermentans, Lb. brevis, Lb. buchneri, Lb. bulgaricus, Lb. carnis, Lb. coryniformis* subsp. *torquens, Lb. curvatus, Lb. delbrueckii, Lb. divergens, Lb. farciminis, Lb. fructosus,*

Lb. hilgardii, Lb. kefir, Lb. lactis subsp. *cremoris, Lb. plantarum,, Lb. paracasei* subsp. *pseudoplantarum, Lb. paracasei* subsp. *paracasei, Lb. salivarius, Lb. sake, Lb. sanfransisco; Lactococcus lactis, Lc. lactis* subsp. *lactis,
Lc. plantarum; Leuconostoc fallax, Leuc. lactis, Leuc. mesenteroides, Leuc.
citreum; Pediococcus pentosaceus, P. acidilactici; Tetragenococcus halophilus; Weisella confuses.*

Bacilli: *Bacillus cereus, B. circulans, B. coagulans, B. laterosporus, B. lentus, B. licheniformis, B. mycoides, B. pumilus, B. sphaericus, B. subtilis,
B. thuringiensis.*

Other bacteria: *Alkaligenes* spp., *Micrococcu*s spp., *Staphylococcus aureus,
S. sciuri.*

Yeasts: *Candida bombicola, C. castellii, C. chiropterorum, C. glabrata, C. famata,
C. parapsilosis, C. montana, C. tropicalis, C. versatilis; Debaryomyces hansenii, D. polymorphus, D. pseudopolymorphus; Geotrichum candidum;
Pichia anomala, P. burtonii, P. fabianii, P. guilliermondi; Saccharomyces
cerevisiae, S. bayanus, S. kluyveri; Saccharomycopsis capsularis, S. cratae-
gensis, S. fibuligera; Zygosaccharomyces rouxii.*

Molds: *Aspergillus oryzae; Mucor circinelloides, M. hiemalis; Rhizopus chinensis, R. oryzae, R. stolonifer* variety *lyococcus.*

Some of these strains possess protective and functional properties,
which render them interesting candidates for use as starter culture(s)
for controlled and optimized production of traditional fermented foods.
Information on the characteristics of the microorganisms isolated from
the ethnic fermented foods definitely enriches the database of microbial
diversity from food ecosystems in the Himalayas. Besides being proposed for starter cultures, some strains can be exploited for production of
enzymes, bioactive compounds and other industrial uses. In fact, a gene
bank for preservation of these strains should be set up in the Himalayas
for further studies.

Microbial diversity, functional foods, cultural foods, income generation, food tourism, and food and nutritional security are the broad aspects
of Himalayan fermented foods (Figure 10.1). The ethnic fermented foods
and beverages are important in the context of Himalayan food ecosystems
as well as in terms of culture, tradition, cost-effectiveness, and nutrition.
Due to cultural adaptation for consumption, and the preserved natures
of the products, most Himalayan fermented foods can be considered for
food and nutritional security of the region. The Himalayan people contribute their indigenous knowledge on the production and management
of available food bioresources (Figure 10.2). This in turn supplements
the food ecosystem and also enhances the regional economy. Based on
this model, some of the food resources can be commercialized in the
Himalayan regions.

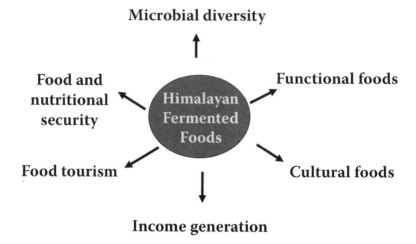

Figure 10.1 Prospects of Himalayan fermented foods.

Figure 10.2 Contribution of indigenous knowledge of the Himalayan people in production and management of ethnic foods.

10.5   Conclusion

Some Himalayan ethnic fermented foods and beverages have medicinal value. Although clinical study of these foods has not yet been done, the Himalayan people strongly believe that they have certain therapeutic values. Such foods, if studied properly, may be projected into the global market. A research and development center for the Himalayan ethnic foods should be set up to open the way for industrialization of some common and functional ethnic food production, with consequent benefits to the regional economy and employment.

The art and culture of making traditional foods at the household level will have to meet the terms of regulatory standards. It is important to link ethnic foods with guidelines for food safety, health, nutrition, and economy. Improvement of ethnic foods through the interface of modern food biotechnology is the need of the day. However, availability and consistent supply of raw materials, basic infrastructure, administrative policy, cost of capital, legislation and trade issues, and import/export restrictions must all be considered before industrializing Himalayan ethnic foods. Genesis of the Himalayan ethnic fermented foods is linked with the cultural civilization of the people, and any proposed improvements should be based on the ethnic demand and cultural interface. One has to respect and recognize the native skills and indigenous expertise of the Himalayan ethnic people for building up a menu of ethnic foods that are worthy of global attention.

Bibliography

Abdel Gawad, A. S. 1993. Effect of domestic processing on oligosaccharide content on some dry legume seeds. *Food Chemistry* 46: 25–31.

Adams, M. R. 1998. Fermented fish. In *Fish and seafood*. Vol. 3 of *Microbiology handbook*, ed. R. A. Lawley and P. Gibbs, 157–177. Leatherhead, Surrey: Leatherhead Food RA.

Adams, M. R. and L. Nicolaides. 1997. Review of the sensitivity of different foodborne pathogens to fermentation. *Food Control* 8 (5, 6): 227–239.

Adams, M. R. and M. J. R. Nout. 2001. *Fermentation and food safety*. Gaithersburg, Md.: Aspen Publishers.

Adhikari, R. R. and H. Ghimirey 2000. *Nepalis society and culture*. Kathmandu, Nepal: Vidharthi Pushtak Bhandar.

Agrahar-Murungkar, D. and G. Subbulakshmi. 2006. Preparation techniques and nutritive value of fermented foods from the Khasi tribes of Meghalaya. *Ecology of Food and Nutrition* 45: 27–38.

Aidoo, K. E., M. J. R. Nout, and P. K. Sarkar. 2006. Occurrence and function of yeasts in Asian indigenous fermented foods. *FEMS Yeast Research* 6: 30–39.

Albrecht, W. J., G. C. Mustakas, and J. E. McGhee. 1966. Rates studies on atmospheric steaming and immersion cooking of soybeans. *Cereal Chemistry* 43: 400.

Ammor, M. S. and B. Mayo. 2007. Selection criteria for lactic acid bacteria to be used as functional starter cultures in dry sausage production: An update. *Meat Science* 76 (1): 138–146.

Ammor, S., G. Tauveron, E. Dufour, and I. Chevallier. 2006. Antibacterial activity of lactic acid bacteria against spoilage and pathogenic bacteria isolated from the same meat small-scale facility; 1: Screening and characterization of the antimicrobial compounds. *Food Science* 17 (6): 454–461.

Ammor, S. M., A. B. Flòrez, and B. Mayo. 2007. Antibiotic resistance in non-enterococcal lactic acid bacteria and bifidobacteria. *Food Microbiology* 24: 559–570.

Amoa, B. and H. S. Muller. 1976. Studies in kenkey with particular reference to calcium and phytic acid. *Cereal Chemistry* 53: 365–375.

Annual Progress Report 2005. *Annual progress report for the year 2004–2005*. Department of Food Security and Agriculture Development, Government of Sikkim, Gangtok, India.

Arora, G., B. H. Lee, and M. Lamoureux. 1990. Characteristics of enzyme profiles of *Lactobacillus casei* species by a rapid API-ZYM system. *Journal of Dairy Sciences* 73: 264–273.

Aryanta, R. W., G. H. Fleet, and K. A. Buckle. 1991. The occurrence and the growth of microorganisms during the fermentation of fish sausage. *International Journal of Food Microbiology* 13: 143–156.

Ashenafi, M. 1989. Proteolytic, lipolytic and fermentative properties of yeasts isolated from ayid, a traditional Ethiopian cottage cheese. *SINET: Ethiopian Journal of Science* 12 (2): 131–139.

Attenborough, R., M. Attenborough, and A. R. Leeds. 1994. Nutrition in *stongde*. In *Himalayan Buddhist villages*, eds. J. Crook and H. Osmaston, 383–404. New Delhi: Motilal Banarsidass.

Axelsson, L. 1998. Lactic acid bacteria: Classification and physiology. In *Lactic acid bacteria: Microbiology and functional aspects*, 2nd ed., eds. S. Salminen and A. V. Wright, 1–72. New York: Marcel Dekker.

Aymerich, T., P. A. Picouet, and J. M. Monfort. 2008. Decontamination technologies for meat products. *Meat Science* 78 (1–2): 114–129.

Backhoff, P. H. 1976. Some chemical changes in fish silage. *Journal of Food Technology* 11: 353–363.

Bacus, J. N. and W. L. Brown. 1981. Use of microbial cultures: Meat products. *Food Technology* 35 (1): 74–78.

Badis, A., D. Guetarni, B. Moussa-Boudjemaa, D. E. Henni, M. E. Tornadijo, and M. Kihal. 2004. Identification of cultivable lactic acid bacteria isolated from Algerian raw goat's milk and evaluation of their technological properties. *Food Microbiology* 21: 343–349.

Bajracharya, D. and T. B. Shrestha. 1973. *Lichhavikalko abhilekh*. Kathmandu, Nepal: Centre for Nepal and Asian Studies.

Balaraman, N. and M. M. Golay. 1991 *Livestock production in Sikkim*. Gangtok, India: Sikkim Science Society.

Balasubramanyam, B. V. and M. C. Vardaraj. 1994. Dahi as a potential source of lactic acid bacteria active against foodborne pathogenic and spoilage bacteria. *Journal of Food Science and Technology* 31 (3): 241–243

Barnett, J. A., R. W. Payne, and D. Yarrow. 1983. *Yeasts: Characteristics and identification*. Cambridge: Cambridge University Press.

Barrangou, R., S.-S. Yoon, F. Breidt Jr., H. P. Fleming, and T. R. Klaenhammer. 2002. Identification and characterization of *Leuconostoc fallax* strains isolated from an industrial sauerkraut fermentation. *Applied and Environmental Microbiology* 68: 2877–2884.

Barthakur, I. K. 1981. Shifting cultivation and economic change in the north-eastern Himalaya. In *The Himalaya: Aspects of change*, ed. J. S. Lall, 447–460. Delhi: Oxford University Press.

Baruzzi, F., A. Matarante, L. Caputo, and M. Marea. 2006. Molecular and physiological characterization of natural microbial communities isolated from a traditional southern Italian processed sausage. *Meat Science* 72 (2): 261–269.

Basappa, S. C. 2002. Investigations on *Chhang* form finger millet (*Eleucine coracana* Gaertn.) and its commercial prospects. *Indian Food Industry* 21 (1): 46–51.

Basappa, S. C., D. Somashekar, R. Agrawal, K. Suma, and K. Bharathi. 1997. Nutritional composition of fermented *ragi* (*chhang*) by *phab* and defined starter cultures as compared to unfermented *ragi* (*Eleucine coracana* G.). *International Journal of Food Science and Nutrition* 48: 313–319.

Basar, K. and N. S. Bisht. 2002. Utilization of bamboo shoots by Adi tribe of Arunachal Pradesh. *Forestry Bull.* 2 (1): 35–57.

Batra, L. R. 1981. Fermented cereals and grain legumes of India and vicinity. In *Advances in biotechnology*. Vol 2, ed. M. Moo-Young and C. W. Robinson, 547–553. New York: Pergamon Press.

Batra, L. R. 1986. Microbiology of some fermented cereals and grains legumes of India and vicinity. In *Indigenous fermented food of non-Western origin*, ed. C. W. Hesseltine and H. L. Wang, 85–104. Berlin: J. Cramer.

Batra, L. R. and P. D. Millner. 1974. Some Asian fermented foods and beverages and associated fungi. *Mycologia* 66: 942–950.

Batra, L. R. and P. D. Millner. 1976. Asian fermented foods and beverages. *Development in Industrial Microbiology* 17: 117–128.

Battcock, M. and S. Azam-Ali. 1998. Fermented fruits and vegetables, a global perspective. FAO Agricultural Services Bulletin, vol. 134. Food and Agriculture Organization, Rome.

Beddows, C. G. 1985. Fermented fish and fish products. In *Microbiology of fermented foods*. Vol. 2, ed. B. J. B. Wood, 1–39. London: Elsevier Applied Science.

Beumer, R. R. 2001. Microbiological hazards and their control: Bacteria. In *Fermentation and food safety*, ed. M. R. Adams and M. J. R. Nout, 141–157. Gaithersburg, Md.: Aspen Publishers.

Bhanja, K. C. 1993. *History of Darjeeling and the Sikkim Himalaya*. New Delhi: Gyan Publishing House.

Bhatia, A. K., R. P. Singh, and C. K. Atal. 1977. Chhang: The fermented beverage of Himalayan folk. *Indian Food Packer* 4: 1–8.

Bhatt, B. P., L. B. Singha, K. Singh, and M. S. Sachan. 2003. Some commercial edible bamboo species of North East India: Production, indigenous uses, cost-benefit and management strategies. *Bamboo Science and Culture* 17 (1): 4–20.

Bhatt, B. P., L. B. Singha, M. S. Sachan, and K. Singh. 2005. Commercial edible bamboo species of the North-Eastern Himalayan region, India. Part 2: Fermented, roasted and boiled bamboo shoot sales. *Journal of Bamboo and Rattan* 4 (1): 13–31.

Bista, D. B. 1967. *People of Nepal*. Kathmandu, Nepal: Ratna Pushtak Bhandar.

Blandino, A., M. E. Al-Aseeri, S. S. Pandiella, D. Cantero, and C. Webb. 2003. Cereal-based fermented foods and beverages. *Food Research International* 36: 527 543.

Bo, T. 1984. Tousi no genryu oyobi sono seisan-gijutu. *Nippon Jouzo Kyoukai Zasshi* 77 (7): 439–445 (Japanese).

Bouton, Y., P. Guyot, and P. Grappin. 1998. Preliminary characterization of microflora of Comte cheese. *Journal of Applied Microbiology* 85: 123–131.

Bover-Cid, S., W. H. and Holzapfel. 1999. Improved screening procedure for biogenic amine production by lactic acid bacteria. *International Journal of Food Microbiology* 53: 33–41.

Brandt, M. J. 2007. Sourdough products for convenient use in baking. *Food Microbiology* 24: 161–164.

Breidt, F., K. A. Crowley, and H. P. Fleming. 1995. Controlling cabbage fermentations with nisin and nisin-resistant *Leuconostoc mesenteroides*. *Food Microbiology* 12 (2): 109–116.

Buchenhüskes, H. J. 1993. Selection criteria for lactic acid bacteria to be used as starter cultures in various food commodities. *FEMS Microbiology Reviews* 12: 253–272.

Campbell-Platt, G. 1987. *Fermented foods of the world: A dictionary and guide.* London: Butterworths.

Campbell-Platt, G. 1994. Fermented foods: A world perspective. *Food Research International* 27: 253–257.

Carr, F. J., D. Chill, and N. Maida. 2002. The lactic acid bacteria: A literature survey. *Critical Review in Microbiology* 28 (4): 281–370.

Cartel, M., M. Erbas, M. K. Uslu, and M. O. Erbas. 2007. Effects of fermentation time and storage on the water-soluble vitamin contents of tarhana. *Journal of Science and Food Agriculture* 87: 1215–1218.

Chammas, G. I., R. Saliba, G. Corrieu, and C. Béal. 2006. Characterisation of lactic acid bacteria isolated from fermented milk "laban." *International Journal of Food Microbiology* 110: 52–61.

Census of India 2001. www.censusindia.gov.in.

Chang, R., S. Schwimmer, and H. K. Burr. 1977. Phytate: Removal from whole dry beans by enzymatic hydrolysis and diffusion. *Journal of Food Science* 42: 1098.

Chaudhuri, H. and S. M. Banerjee. 1965. Report on the fisheries of Manipur, with special reference to the development of the Takmu Beel area of Loktak Lake. *Miscellaneous Contr., Central Inland Fishery Research Institute, Barrackpore* 4: 1–29.

Chavan, J. K. and S. S. Kadam. 1989. Nutritional improvement of cereals by fermentation. *Critical Reviews in Food Science and Nutrition* 28: 349–400.

Cheigh, H.-S. and K.-Y. Park. 1994. Biochemical, microbiological, and nutritional aspects of kimchi (Korean fermented vegetable products). *Critical Reviews in Food Science and Nutrition* 34 (2): 175–203.

Chettri, R. and J. P. Tamang. 2008. Microbiological evaluation of *maseura*, an ethnic fermented legume-based condiment of Sikkim. *Journal of Hill Research* 21 (1): 1–7.

Chitale, S. R. 2000. Commercialization of Indian traditional foods: *Jeelebi, laddoo* and *bakervadi.* In *Proceedings of the International Conference on Traditional Foods,* CFTRI, 331. March 6–8, 1997, Central Food Technological Research Institute, Mysore, India.

Cibik, R., E. Lepage, and P. Tailliez. 2000. Molecular diversity of *Leuconostoc mesenteroides* and *Leuconostoc citreum* isolated from traditional French cheese as revealed by RAPD fingerprinting, 16S rDNA sequencing and 16S rDNA fragment amplification. *Systematic and Applied Microbiology* 23: 267–278.

Cintas, L. M., P. Cassaus, C. Herranz, L. S. Havarstein, H. Holo, P. Hernandez, and I. F. Nes. 2000. Biochemical and genetic evidence that *Enterococcus faecium* L50 produces enterocins L50A and L50B, the sec-dependent enterocin P, and a novel bacteriocin secreted without an N-terminal extension termed enterocin Q. *Journal of Bacteriology* 182: 6806–6814.

Claus, D. and R. C. W. Berkeley. 1986. Genus *Bacillus* Cohn 1872, 174. In *Bergey's manual of systematic bacteriology.* Vol. 2, ed. P. H. A. Sneath, N. S. Mair, M. E. Sharpe, and J. G. Holt, 1105–1139. Baltimore, Md.: Williams and Wilkins.

Crisan, E. V. and A. Sands. 1975. Microflora of four fermented fish sauces. *Applied Microbiology* 29: 106–108.

Cronk, T. C., K. H. Steinkraus, L. R. Hackler, and L. R. Mattick. 1977. Indonesia *tapé ketan* fermentation. *Applied and Environmental Microbiology* 33: 1067–1073.

Daeschel, M. A. and H. P. Fleming. 1987. Achieving pure culture cucumber fermentations: A review. In *Developments in industrial microbiology.* Vol. 28, ed. G. Pierce, 141–148. Arlington, Va.: Society for Industrial Microbiology.

Dahal, N., E. R. Rao, and B. Swamylingappa. 2003. Biochemical and nutritional evaluation of *masyaura*: A legume based traditional savoury of Nepal. *Journal of Food Science and Technology* 40: 17–22.

Dahal, N., T. B. Karki, B. Swamylingappa, Q. Li, and G. Gu. 2005. Traditional foods and beverages of Nepal: A review. *Food Review International* 21: 1–25.

Das, C. P. and A. Pandey. 2007. Fermentation of traditional beverages prepared by Bhotiya community of Uttaranchal Himalaya. *Indian Journal of Traditional Knowledge* 6 (1): 136–140.

Davies, F. L. and B. A. Law. 1984. *Advances in the microbiology and biochemistry of cheese and fermented milk*. New York: Elsevier Applied Science.

Deak, T. 1991. Food borne yeast. *Advances in Applied Microbiology* 36: 179–278.

Decock, P. and S. Cappelle. 2005. Bread technology and sourdough technology. *Trends in Food Science and Technology* 16: 113–120.

Delgado, A., D. Brito, P. Fevereiro, R. Tenreiro, and C. Peres. 2005. Bioactivity quantification of crude bacteriocin solutions. *Journal of Microbiological Methods* 62 (1): 121–124.

de Man, J. C., M. Rogosa, and M. E. Sharpe. 1960. A medium for the cultivation of lactobacilli. *Journal of Applied Bacteriology* 23: 130–135.

Demeyer, D., M. Raemaekers, A. Rizzo, A. Holck, A. De Smedt, B. ten Brink, B. Hagen, C. Montel, E. Zanardi, E. Murbrekk, F. Leroy, F. Vandendriessche, K. Lorentsen, L. H. Sunesen, L. De Vuyst, R. Talon, R. Chizzolini, and S. Eerola. 2000. Control of bioflavour and safety of fermented sausage: First results of a European project. *Food Research International* 33: 171–180.

de Vuyst, L. 2000. Technology aspects related to the application of functional starter culture. *Food Technology and Biotechnology* 38 (2): 105–112.

Dewan, S. 2002. Microbiological evaluation of indigenous fermented milk products of the Sikkim Himalayas. Ph.D. thesis, Food Microbiology Laboratory, Sikkim Government College (under North Bengal University), Gangtok, India.

Dewan, S. and J. P. Tamang. 2006. Microbial and analytical characterization of Chhu, a traditional fermented milk product of the Sikkim Himalayas. *Journal of Scientific and Industrial Research* 65: 747–752.

Dewan, S. and J. P. Tamang. 2007. Dominant lactic acid bacteria and their technological properties isolated from the Himalayan ethnic fermented milk products. *Antonie van Leeuwenhoek International Journal of General and Molecular Microbiology* 92 (3): 343–352.

Dhavises, G. 1972. Microbial studies during the pickling of the shoot of bamboo, *Bambusa arundinacea* Willd., and of *pak sian*, *Gynandropsis pentaphylla* D.C. M.S. thesis, Kasetsart University, Bangkok.

Dietz, H. M. 1984. Fermented dried vegetables and their role in nutrition in Nepal. *Proceeding of Institute of Food Science and Technology, UK* 17: 208–213.

Ding, H. and C. Lämmler. 1992. Cell surface hydrophobicity of *Actinomyces pyogenes* determined by hexadecane adherence and salt aggregation studies. *Zentralblatt für Veterinärmedizin (B)* 39: 132–138.

Dung, N. T. P., F. M. Rombouts, and M. J. R. Nout. 2006a. Development of defined mixed-culture fungal fermentation starter granulate for controlled production of rice wine. *Innovative Food Science and Emerging Technologies* 6: 429–441.

Dung, N. T. P., F. M. Rombouts, and M. J. R. Nout. 2006b. Functionality of selected strains of moulds and yeasts from Vietnamese rice wine starters. *Food Microbiology* 23: 331–340.

Durlu-Ozkaya, F., V. Xanthopoulos, N. Tunaï, and E. Litopoulou-Tzanetaki. 2001. Technologically important properties of lactic acid bacteria isolated from Beyaz cheese made from raw ewes' milk. *Journal of Applied Microbiology* 91: 861–870.

du Toit, M., C. M. A. P. Franz, L. M. T. Dicks, U. Schillinger, P. Haberer, B. Warlies, F. Ahrens, and W. H. Holzapfel. 1998. Characterisation and selection of probiotic lactobacilli for a preliminary minipig feeding trial and their effect on serum cholesterol levels, faeces pH and faeces moisture content. *International Journal of Food Microbiology* 40: 93–104.

Dykes, G. A., T. J. Britz, and A. von Holy. 1994. Numerical taxonomy and identification of lactic acid bacteria from spoiled, vacuum packaged Vienna sausages. *Journal of Applied Bacteriology* 76: 246–252.

Dzudie, T., M. Bouba, C. M. Mbofung, and J. Scher. 2003. Effect of salt dose on the quality of dry smoked beef. *Italian Journal of Food Science* 15: 433–440.

Efiuvwevwere, B. J. O. and O. Akona. 1995. The microbiology of *kununzaki*, a cereal beverage from northern Nigeria, during the fermentation (production) process. *World Journal of Microbiology and Biotechnology* 11: 491–493.

Efiuvwevwere, B. J. O. and C. F. Ezeama. 1996. Influence of fermentation time and an indigenous tenderiser (kanwa) on the microbial profile, chemical attributes and shelf-life of rice *masa* (a Nigerian fermented product). *Journal of Science of Food and Agriculture* 71: 442–448.

Eom, H.-J., D. M. Seo, and N. S. Han. 2007. Selection of psychrotrophic *Leuconostoc* spp. producing highly active dextransucrase from lactate fermented vegetables. *International Journal of Food Microbiology* 117 (1): 61–67.

Erbas, M., M. Certel, and M. K. Uslu. 2005a. Microbial and chemical properties of *tarhana* during fermentation and storage as wet-sensorial properties of *tarhana* soup. *LWT* 38: 409–416.

Erbas, M., M. F. Ertugay, M. O. Erbas, and M. Certel. 2005b. The effect of fermentation and storage on free amino acids of *tarhana*. *International Journal of Food Sciences and Nutrition* 56 (5): 349–358.

Erbas, M., M. K. Uslu, M. O. Erbas, and M. Certel. 2006. Effects of fermentation and storage on the organic and fatty acid contents of *tarhana*, Turkish fermented cereal food. *Journal of Food Composition and Analysis* 19: 294–301.

Etchells, J. L., T. A. Bell, H. P. Fleming, R. E. Kelling, and R. L. Thompson. 1973. Suggested procedure for the controlled fermentation of commercially brined pickling cucumbers: The use of starter cultures and reduction of carbon dioxide accumulation. *Pickle Pak Science* 3: 4–14.

Farnworth, E. R. 2003. *Handbook of fermented functional foods*. Food Research and Development Centre, Agriculture and Agri-Food Canada, 251–275. Boca Raton, Fla.: CRC Press.

Fernandez Gonzalez, M. J., P. Garcia Garcia, A. Garrido Fernandez, and M. C. Duran Quintana. 1993. Microflora of the aerobic preservation of directly brined green olives from Hojiblanca cultivar. *Journal of Applied Bacteriology* 75 (3): 226–233.

Ferreira, V., J. Barbosa, S. Vendeiro, A. Mota, F. Silva, M. J. Monteiro, T. Hogg, P. Gibbs, and P. Teixeira. 2006. Chemical and microbiological characterization of *alheira*: A typical Portuguese fermented sausage with particular reference to factors relating to food safety. *Meat Science* 73: 570–575.

Fleming, H. P. 1984. Developments in cucumber pickling fermentation. *Journal of Chemical Technology and Biotechnology* 34B: 241–252.

Fleming, H. P. and R. F. McFeeters. 1981. Use of microbial cultures: Vegetable products. *Food Technology* 35: 84–88.

Fleming, H. P., R. F. McFeeters, and M. A. Daeschel. 1985. The lactobacilli, pediococci and leuconostocs: Vegetable products. In *Bacterial starter cultures for foods*, ed. S. E. Gilliland, 97–115. Boca Raton, Fla.: CRC Press.

Franz, C. M. A. P., M. E. Stiles, K. H. Schleifer, and W. H. Holzapfel. 2003. Enterococci in foods: A conundrum for food safety. *International Journal of Food Microbiology* 88: 105–122.

Fredrikson, M., T. Andlid, A. Haikara, and A. S. Sandberg. 2002. Phytate degradation by micro-organisms in synthetic media and pea flour. *Journal of Applied Microbiology* 93: 197–204.

Fujii, T., M. Matsubara, Y. Itoh, and M. Okuzumi. 1994. Microbial contributions on ripening of squid shiokara. *Nippon Suisan Gakkaishi* 60: 265–270.

Gadaga, T. H., A. N. Mutukumira, and J. A. Narhus. 2001. Growth characteristics of *Candida kefyr* and two strains of *Lactococcus lactis* subsp. *lactis* isolated from Zimbabwean naturally fermented milk. *International Journal of Food Microbiology* 70: 11–19.

Gaya, P., M. Babin, M. Medina, and M. Nunez. 1999. Diversity among lactococci isolated from ewes' raw milk and cheese. *Journal of Applied Microbiology* 87: 849–855.

Geisen, R. and W. H. Holzapfel. 1996. Genetically modified starters and protective cultures. *International Journal of Food Microbiology* 30: 315–324.

Giri, S. S. and L. S. Janmejay. 1987. Microbial and chemical contents of the fermented bamboo shoot *soibum*. *Frontier Botany* 1: 89–100.

Giri, S. S. and L. S. Janmejay. 1994. Changes in soluble sugars and other constituents of bamboo shoots in *soibum* fermentation. *Journal of Food Science and Technology* 31 (6): 500–502.

Giri, S. S. and L. S. Janmejay. 2000. Effects of bamboo shoot fermentation and aging on nutritional and sensory qualities of *soibum*. *Journal of Food Science and Technology* 37 (4): 423–426.

Gode, P. K. 1943. Some notes on the history of Indian dietetics with special reference to the history of *jalebi*. *New Indian Antiquary* 6: 169–181.

Gopalan, C., B. V. Rama Sastri, S. C. Balasubramanian, B. S. Narasinga Rao, Y. G. Deosthale, and K. C. Pant. 2004. *Nutritive value of Indian foods* (revised edition). Hyderabad, India: National Institute of Nutrition.

Gordon, R. E., W. C. Haynes, and C. H.-N. Pang. 1973. The genus *Bacillus*, Handbook No. 427. United States Department of Agriculture, Washington, D.C.

Gorer, G. 1938. *The Lepchas of Sikkim*. Delhi: Gian Publishing House.

Gran, H. M., H. T. Gadaga, and J. A. Narvhus. 2003. Utilisation of various starter cultures in the production of *amasi*, a Zimbabwean naturally fermented raw milk product. *International Journal of Food Microbiology* 88: 19–28.

Grant, I. R. and M. F. Patterson. 1991. A numerical taxonomic study of lactic acid bacteria isolated from irradiated pork and chicken packaged under various gas atmospheres. *Journal of Applied Bacteriology* 70: 302–307.

Gupta, M., N. Khetarpaul, and B. M. Chauhan. 1992a. *Rabadi* fermentation of wheat: Changes in phytic acid content and in vitro digestibility. *Plant Foods for Human Nutrition* 42: 109–116.

Gupta, M., N. Khetarpaul, and B. M. Chauhan. 1992b. Preparation, nutritional value and acceptability of barley *rabadi*: An indigenous fermented food of India. *Plant Foods for Human Nutrition* 42: 351–358.

Haard, N. F., A. A. Odunfa, C. H. Lee, R. Quintero-Ramírez, A. Lorence-Quiñones, and C. Wacher-Radarte. 1999. Fermented cereals: A global perspective. FAO Agricultural Service Bulletin, vol. 138, 63–97. Food and Agriculture Organization, Rome.

Hadisepoetro, E. S. S., N. Takada, and Y. Oshima. 1979. Microflora in *ragi* and *usar*. *Journal of Fermentation Technology* 57: 251–259.

Halász, A., A. Baráth, L. Simon-Sarkadi, and W. H. Holzapfel. 1994. Biogenic amines and their production by microorganisms in food. *Trends in Food Science and Technology* 5: 42–49.

Hall, C. M., L. Sharles, R. Mitchell, N. Macionis, and B. Cambourne. 2003. *Food tourism around the world*. London: Butterworth Heinemann.

Hammes, W. P. and M. G. Ganzle. 1998. Sourdough breads and related products. In *Microbiology of fermented foods,* 2nd ed., ed. B. J. B. Wood, 199–216. Glasgow: Blackie Academic and Professional.

Hammes, W. P. and R. H. Vogel. 1995. The genus *Lactobacillus*. In *The genera of lactic acid bacteria*. Vol. 2 of *The lactic acid bacteria*, ed. B. J. B. Wood and W. H. Holzapfel, 19–54. London: Blackie Academic & Professional, an Imprint of Chapman and Hall.

Hammes, W. P., M. J. Brandt, K. L. Francis, U. J. Rosenheim, F. H. Seitter, and A. Vogelmann. 2005. Microbial ecology of cereal fermentations. *Trends in Food Science and Technology* 16: 4–11.

Han, B. Z., R. R. Beumer, F. M. Rombouts, and M. J. R. Nout. 2001. Microbiological safety and quality of commercial *sufu*: A Chinese fermented soybean food. *Food Control* 12: 541–547.

Hansen, E. B. 2002. Commercial bacterial starter cultures for fermented foods of the future. *International Journal of Food Microbiology* 78: 119–131.

Hara, T., S. Hiroyuki, I. Nobuhide, and K. Shinji. 1995. Plasmid analysis in polyglutamate-producing *Bacillus* strain isolated from non-salty fermented soybean food, *kinema*, in Nepal. *Journal of General and Applied Microbiology* 41: 3–9.

Haridas Rao, P. 2000. Developments in wheat based Indian traditional foods. In *Proceedings of the 1997 International Conference on Traditional Foods,* CFTRI, 276–284, March 6–8, 1997, CFTRI, Mysore, India.

Harris, L. J. 1998. The microbiology of vegetable fermentations. In *Microbiology of fermented foods*. Vol. 1, ed. B. J. B. Wood, 45–72. London: Blackie Academic and Professional.

Hastings, J. W. and W. H. Holzapfel. 1987a. Conventional taxonomy of lactobacilli surviving radurization of meat. *Journal of Applied Bacteriology* 62: 209–216.

Hastings, J. W. and W. H. Holzapfel. 1987b. Numerical taxonomy of lactobacilli surviving radurization of meat. *International Journal of Food Microbiology* 4: 33–49.

Herreros, M. A., J. M. Fresno, M. J. González Prieto, and M. E. Tornadijo. 2003. Technological characterisation of lactic acid bacteria isolated from Armada cheese (a Spanish goats' milk cheese). *International Dairy Journal* 13: 469–479.

Hesseltine, C. W. 1965. A millennium of fungi, food and fermentation. *Mycologia* 57: 149–197.

Hesseltine, C. W. 1979. Some important fermented foods of Mid-Asia, the Middle East, and Africa. *Journal of American Oil Chemists' Society* 56: 367–374.

Hesseltine, C. W. 1983. Microbiology of Oriental fermented foods. *Annual Review of Microbiology* 37: 575–601.

Hesseltine, C. W. 1985a. Genus *Rhizopus* and *tempeh* microorganisms. In *Proceedings of the Asian symposium of non-salted soybean fermentation*, July 1985. National Food Research Institute, Tsukaba Science City, Tsukaba, Japan, 20–21.

Hesseltine, C. W. 1985b. Fungi, people, and soybeans. *Mycologia* 77: 505–525.

Hesseltine, C. W. 1991. Zygomycetes in food fermentations. *Mycologist* 5 (4): 162–169.

Hesseltine, C. W. and C. P. Kurtzman. 1990. Yeasts in amylolytic food starters. *Anales del Instituto de Biologia, Universidad Nacional Autónoma de México, Serie Botanica* 60: 1–7.

Hesseltine, C. W. and M. L. Ray. 1988. Lactic acid bacteria in *murcha* and *ragi*. *Journal of Applied Bacteriology* 64: 395–401.

Hesseltine, C. W. and H. L. Wang. 1967. Traditional fermented foods. *Biotechnology and Bioengineering* 9: 275–288.

Hesseltine, C. W., R. Rogers, and F. G. Winarno. 1988. Microbiological studies on amylolytic Oriental fermentation starters. *Mycopathologia* 101: 141–155.

Hitchcock, J. T. and R. L. Jones. 1994. *Spirit possession in the Nepal Himalayas*. New Delhi: Vikas Publishing.

Holzapfel, W. H. 1997. Use of starter cultures in fermentation on a household scale. *Food Control* 8 (5–6): 241–258.

Holzapfel, W. H. 2002. Appropriate starter culture technologies for small-scale fermentation in developing countries. *International Journal of Food Microbiology* 75: 197–212.

Holzapfel, W. H. and U. Schillinger. 2002. Introduction to pre- and probiotics. *Food Research International* 35: 109–116.

Holzapfel, W. H., R. Giesen, and U. Schillinger. 1995. Biological preservation of foods with reference to protective cultures, bacteriocins and food-grade enzymes. *International Journal of Food Microbiology* 24: 343–362.

Holzapfel, W. H., P. Haberer, J. Snel, U. Schillinger, and J. H. J. Huis in't Veld. 1998. Overview of gut flora and probiotics. *International Journal of Food Microbiology* 41: 85–101.

Holzapfel, W. H., U. Schillinger, M. D. Toit, and L. Dicks. 1997. Systematics of probiotic lactic acid bacteria with reference to modern phenotypic and genomic methods. *Microecology and Therapy* 26: 1–10.

Holzapfel, W. H., U. Schillinger, R. Giesen, and F.-K. Lücke. 2003. Starter and protective cultures. In *Food preservatives*, ed. N. J. Russell and G. W. Gould, 291–319. 2nd ed. New York: Kluwer Academic/Plenum Publishers.

Hooker, J. D. 1854. *Himalayan journals: Notes of a naturalist in Bengal, the Sikkim and Nepal Himalayas, the Khasia Mountains*. London: John Murray.

Hore, D. K. 1998. Genetic resources among bamboos of North East India. *Journal of Economic and Taxonomic Botany* 22 (1): 173–181.

Hosoi, T. and K. Kiuchi. 2003. *Natto*: A food made by fermenting cooked soybeans with *Bacillus subtilis* (natto). In *Handbook of fermented functional foods*, ed. E. R. Farnworth, 227–250, New York: CRC Press.

Hossain, S. A., P. K. Pal, P. K. Sarkar, and G. R. Patil. 1996. Sensory characteristics, manufacturing methods and cost of producing of milk *churpi*. *Journal of Hill Research* 9 (1): 121–127.

Hounhouigan, D. J., M. J. R. Nout, C. M. Nago, J. H. Houben, and F. M. Rombouts. 1993a. Changes in the physico-chemical properties of maize during natural fermentation of *mawè*. *Journal of Cereal Science* 17: 291–300.

Hounhouigan, D. J., M. J. R. Nout, C. M. Nago, J. H. Houben, and F. M. Rombouts. 1993b. Characterization and frequency distribution of species of lactic acid bacteria involved in the processing of *mawè*, a fermented maize dough from Benin. *International Journal of Food Microbiology* 18: 279–287.

Hounhouigan, D. J., M. J. R. Nout, C. M. Nago, J. H. Houben, and F. M. Rombouts. 1993c. Composition and microbiological and physical attributes of *mawè*, a fermented maize dough from Benin. *International Journal of Food Microbiology* 28: 513–517.

Hulse, J. H. 2004. Biotechnologies: Past history, present state and future prospects. *Trends in Food Science* 15: 3–18.

Hutchinson, C. M. and C. S. Ram-Ayyar. 1925. *Bakhar*, the Indian rice beer ferment. *Memoirs of the Department of Agriculture in India, Bacteriology Series* 1: 137–168.

Hymowitz, T. 1970. On the domestication of the soybean. *Economic Botany* 24 (4): 408–421.

Hymowitz, T. and N. Kaizuma. 1981. Soybean seed protein electrophoresis profiles from 15 Asian countries or regions: Hypothesis on paths of dissemination of soybeans from China. *Economic Botany* 35 (1): 10–23.

Igoumenidou, V., I. Lambropoulos, I. Roussou, and I. G. Roussis. 2005. Casein and cheese peptide degradation by *Enterococcus durans* FC12 isolated from Feta cheese. *Food Biotechnology* 19 (2): 161–172.

Inatsu, Y., N. Nakamura, Y. Yuriko, T. Fushimi, L. Watanasiritum, and S. Kawamoto. 2006. Characterization of *Bacillus subtilis* strains in *Thua nao*, a traditional fermented soybean food in northern Thailand. *Letters in Applied Microbiology* 43: 237.

Inoue, T., J. Tanaka, and S. Mitsui. 1992. *Recent advances in Japanese brewing technology*, Vol. 2 (1). Tokyo: Gordon and Breach Science Publishers.

Ishige, N. 1993. Cultural aspects of fermented fish products in Asia. In *Fish fermentation technology*, ed. C. H. Lee, K. H. Steinkraus, and P. J. A. Reilly, 13–32. Tokyo: United Nations University Press.

Itabashi, M. 1986. *Sunki*-pickles prepared by single species of lactic acid bacteria. *Nippon Shokuhin Kogyo Gakkaishi* 34 (6): 356–361.

Ito, H., J. Tong, and Y. Li. 1996. Chinese *dauchi*, from *itohiki* natto to nonmashed miso. *Miso Science Technology* 44: 224–250.

Ives, J. D. 2006. *Himalayan perception: Environmental change and the well-being of mountain peoples*. London: Routledge.

Ives, J. D. and B. Messerli. 1989. *The Himalayan dilemma: Reconciling development and conservation*. London: Routledge.

Jamir, N. S. and R. R. Rao. 1990. Fifty new or interesting medicinal plants used by the *Zeliang* of Nagaland (India). *Ethnobotany* 2: 11–18.

Jamuna, M., S. T. Babusha, and K. Jeevaratnam. 2005. Inhibitory efficacy of nisin and bacteriocins from *Lactobacillus* isolates against food spoilage and pathogenic organisms in model and food systems. *Food Microbiology* 22 (5): 449–454.

Jenson, I. 1998. Bread and bakers yeasts. In *Microbiology of fermented foods*, ed. B. J. B. Wood, 172–198. 2nd ed. Glasgow: Blackie Academic and Pistermond.

Jeyaram, K., W. Mohendro Singh, T. Premarani, A. Ranjita Devi, K. Selina Chanu, N. C. Talukdar, and M. Rohinikumar Singh. 2008a. Molecular identification of dominant microflora associated with *hawaijar*: A traditional fermented soybean (*Glycine max* [L.]) food of Manipur, India. *International Journal of Food Microbiology* 122: 259–268.

Jeyaram, K., W. Mohendro Singh, A. Capece, and P. Romano. 2008b. Molecular identification of yeast species associated with *hamei*: A traditional starter used for rice wine production in Manipur, India. *International Journal of Food Microbiology* 124: 115–125.

Jhingran, V. G. 1977. *Fish and fisheries of India*. Delhi: Hindustan Publishing.

Joosten, H. M. L. J. and M. D. Northolt. 1989. Detection, growth and amine-producing capacity of lactobacilli in cheese. *Applied Environmental Microbiology* 55: 2356–2359.

Joshi, V. K. and D. K. Sandhu. 2000. Quality evaluation of naturally fermented alcoholic beverages, microbiological examination of source of fermentation and ethanol productivity of the isolates. *Acta Alimentaria* 29 (4): 232–334.

Kandler, O. 1983. Carbohydrate metabolism in lactic acid bacteria. *Antonie van Leeuwenhoek* 49: 209–224.

Kanno, A., T. Haruki, N. Takano, and T. Akimoto. 1982. Change of saccharides in soybean during manufacturing of natto. *Nippon Shikuhin Kogyo Gakkaishi* 29, 105–110.

Kanwar, S. S., M. K. Gupta, C. Katoch, R. Kumar, and P. Kanwar. 2007. Traditional fermented foods of Lahaul and Spiti area of Himachal Pradesh. *Indian Journal of Traditional Knowledge* 6 (1): 42–45.

Karki, T. 1986. Some Nepalese fermented foods and beverages. In *Traditional foods: Some products and technologies*, CFTRI, 84-96. Central Food Technological Research Institute, Mysore.

Karki, T. 1994. Food processing industries in Nepal. In *Proceedings of the International Workshop on Application and Control of Microorganisms in Asia*, ed. K. Komogata, T. Yoshida, T. Nakase, and H. Osada, 71–81. The Institute of Physical and Chemical Research and Japan International Science and Technology Exchange Centre, Japan.

Karki, T., H. Itoh, and M. Kozaki. 1983a. Chemical changes occurring during *gundruk* fermentation: Part 2-2, Flavour components. *Lebensmittel-Wissenschaft und-Technologie* 16: 203–208.

Karki, T., H. Itoh, K. Hayashi, and M. Kozaki. 1983b. Chemical changes occurring during *gundruk* fermentation: Part 2-1, Amino acids. *Lebensmittel-Wissenschaft und-Technologie* 16: 180–183.

Karki, T., H. Itoh, K. Kiuchi, H. Ebine, and M. Kozaki. 1983c. Lipids in *gundruk* and *takana* fermented vegetables. *Lebensmittel-Wissenschaft und-Technologie* 16: 167–171.

Karki, T., S. Okada, T. Baba, H. Itoh, and M. Kozaki. 1983d. Studies on the microflora of Nepalese pickles *gundruk*. *Nippon Shokuhin Kogyo Gakkaishi* 30: 357–367.

Katiyar, S. K., N. Kumar, and A. K. Bhatia. 1989. Traditional milk products of Ladakh tribes. *Arogya Journal of Health Science* 15: 49–52.

Katiyar, S. K., A. K. Bhasin, and A. K. Bhatia. 1991. Traditionally processed and preserved milk products of Sikkimese tribes. *Science and Culture* 57 (10–11): 256–258.

Kawamura, Y. and M. R. Kara. 1987. *Umami: A basic taste*. New York: Marcel Dekker.

KC, J. B., D. K. Subba, and B. K. Rai. 2001. Plants used in *murcha* preparation in eastern Nepal. *Journal of Hill Research* 14 (2): 107–109.

Kelly, W. J., R. V. Asmundson, G. L. Harrison, and C. M. Huang. 1995. Differentiation of dextran-producing *Leuconostoc* strains from fermented rice cake (*puto*) using pulse field electrophoresis. *International Journal of Food Microbiology* 26: 345–352.

Khatri, P. K. 1987. *Nepali Samaj ra Sanskriti (Prachin-Madhyakal)*. Kathmandu, Nepal: Shaja Prakashan.

Khawas, V. 2008. Environmental issues and human security in the Himalayas: A comparative study of Sikkim and Eastern Nepal. Ph.D. thesis, Jawaharlal Nehru University, New Delhi.

Kiers, J. L., A. E. A. Van Laeken, F. M. Rombouts, and M. J. R. Nout. 2000. In vitro digestibility of *Bacillus* fermented soya bean. *International Journal of Food Microbiology* 60: 163–169.

Kim, C. J. 1968. Microbiological and enzymological studies on *takju* brewing. *Journal of Korean Agricultural and Chemical Society* 10: 69–99.

Kimura, S. 2000. Food culture of the 21st century. *Food Culture* 1: 4–5.

Kiuchi, K. 2001. Miso and natto. *Food Culture* 3: 7–10.

Kiuchi, K., T. Ohta, H. Itoh, T. Takabayashi, and H. Ebine. 1976. Studies on lipids of natto. *Journal of Agricultural Food Chemistry* 24: 404–407.

Ko, S. D. 1972. *Tapé* fermentation. *Applied Microbiology* 23: 976–978.

Kobayashi, Y., K. Tubaki, and M. Soneda. 1961. Several moulds and a yeast used for brewing native beer (*kodok jar*) among the Sikkimese of India. *Journal of Japanese Botany* 36: 321–331.

Kostinek, M., I. Specht, V. A. Edward, U. Schillinger, C. Hertel, W. H. Holzpafel, and C. M. A. P. Franz. 2005. Diversity and technological properties of predominant lactic acid bacteria from fermented cassava used for the preparation of *gari*, a traditional African food. *Systematic and Applied Microbiology* 28 (6): 527–540.

Kozaki, M. and T. Uchimura. 1990. Microorganisms in Chinese starter *bubod* and rice wine *tapuy* in the Philippines. *Journal of Brewing Society of Japan* 85 (11): 818–824.

Kozaki, M., J. P. Tamang, J. Kataoka, S. Yamanaka, and S. Yoshida. 2000. Cereal wine (*jaanr*) and distilled wine (*raksi*) in Sikkim. *Journal of Brewing Society of Japan* 95 (2): 115–122.

Kreger-van Rij, N. J. W. 1984. *The yeasts: A taxonomic study*. Amsterdam: Elsevier Science.

Kurtzman, C. P. and J. W. Fell. 1998. *The yeasts: A taxonomic study*. 4th ed. Amsterdam: Elsevier Science.

Kwon, T. W. 1994. The role of fermentation technology for the world food supply. In *Lactic acid fermentation of non-dairy food and beverages*, ed. C. H. Lee, J. Adler-Nissen, and G. Bärwald, 1–7. Seoul: Ham Lim Won.

Latorre-Moratalla, M. L., S. Bover-Cid, T. Aymerich, B. Marcos, M. C. Vodal-Carou, and M. Garrig. 2007. Aminogenesis control in fermented sausages manufactured with pressurized meat batter and starter culture. *Meat Science* 75 (3): 460–469.

Laufer, B. 1914. Some fundamental ideas of Chinese culture. *Journal of Race Development* 5: 160–174.

Lee, A. C. and Y. Fujio. 1999. Microflora of *banh men*, a fermentation starter from Vietnam. *World Journal Microbiology and Biotechnology* 15: 57–62.

Lee, C.-H. 1994. Importance of lactic acid bacteria in non-dairy food fermentation. In *Lactic acid fermentation of non-dairy food and beverages*, ed. C.-H. Lee, J. Adler-Nissen, and G. Bärwald, 8–25. Seoul: Harn Lim Won.

Lee, C.-H. 1995. An introduction to Korean food culture. *Korean and Korean American Studies Bulletin* 6 (1): 6–10.

Lee, C.-H. 1997. Lactic acid fermented foods and their benefits in Asia. *Food Control* 8: 259–269.

Lee, C.-H. and G. M. Kim. 1993. Korean rice-wine: The types and processing methods in old Korean literature. *Bioindustry* 6 (4): 8–25.

Lee, C.-H., K. H. Steinkraus, and P. J. Alan Reilly. 1993. *Fish fermentation technology.* Tokyo: United Nations University Press.

Lee, S. W. 1984. *Hankuk sikpum munhwasa* (Korean dietary culture). Seoul: Kyomunsa (Korean).

Legan, J. D. and P. A. Voysey. 1991. Yeasts spoilage of bakery products and ingredients. *Journal of Applied Bacteriology* 70: 361–371.

Leisner, J. J., J. C. Millan, H. H. Huss, and L. M. Larsen. 1994. Production of histamine and tyramine by lactic acid bacteria isolated from vacuum-packed sugar-salted fish. *Journal of Applied Bacteriology* 76: 417–423.

Limtong, S., S. Sintara, P. Suwannarit, and N. Lotong. 2002. Yeast diversity in Thai traditional alcoholic starter. *Kasetsart Journal of Nutrition Science* 36: 149–158.

Lindgren, S. E. and W. J. Dobrogosz. 1990. Antagonistic activities of lactic acid bacteria in food and feed fermentations. *FEMS Microbiology Review* 87: 149–164.

Liong, M. T. 2008. Safety of probiotics: Translocation and infection. *Nutrition Reviews* 66: 192–202.

Logan, N. A. and R. C. W. Berkeley. 1984. Identification of *Bacillus* strains using the API system. *Journal of General Microbiology* 130: 1871–1882.

Lu, Z., H. P. Fleming, and R. F. McFeeters. 2001. Differential glucose and fructose utilization during cucumber juice fermentation. *Journal of Food Science* 66 (1): 162–166.

Lücke, F. K. 2003. Fermented meat products. In *Encyclopedia of food sciences and nutrition*, ed. B. Carballero, L. C. Trugo, and P. M. Finglas, 2338–2343. 2nd ed. London: Elsevier Science Ltd., Academic Press.

Mabesa, R. C., M. M. Castillo, S. V. Revilla, and V. T. Bandian. 1983. Safety evaluation of fermented fish and shellfish products: I, Microbiological contaminants. *Philippines Journal of Science* 112 (1–2): 91–102.

Mao, A. A. 1998. Ethnobotanical observation of rice beer *zhuchu* preparation by the Mao Naga tribe from Manipur (India). *Bulletin Botanical Survey of India* 40 (1–4): 53–57.

Mao, A. A. and Odyuo, N. 2007. Traditional fermented foods of the *Naga* tribes of northeastern India. *Indian Journal of Traditional Knowledge* 6 (1): 37–41.

Mathara, J. M., U. Schillinger, P. M. Kutima, S. K. Mbugua, and W. H. Holzapfel. 2004. Isolation, identification and characterization of the dominant microorganisms of *kule naoto*: The Maasai traditional fermented milk in Kenya. *International Journal of Food Microbiology* 94: 269–278.

Mbugua, S. K., R. H. Ahrens, H. N. Kigutha, and V. Subramanian. 1992. Effect of fermentation, malted flour treatment and drum drying on nutritional quality of *uji*. *Ecology of Food Nutrition* 28: 271–277.

McDonald, L. C., H. P. Fleming, and M. A. Daeschel. 1991. Acidification effects on microbial populations during initiation of cucumber fermentation. *Journal of Food Science* 56 (5): 1353–1359.

McWilliams, M. 2007. *Food around the world: A cultural perspective.* New Delhi: Pearson Education.

Mensah, P. P. A., A. M. Tomkins, B. S. Drasar, and T. J. Harrison. 1990. Fermentation of cereals for reduction of contamination of weaning foods in Ghana. *The Lancet* 336: 140–143.

Mingmuang, M. 1974. Microbial study during the pickling of leaves of mustard (*Brassica juncea* L.) by fermentation process. M.S. thesis, Kasetsart University, Bangkok.

Misra, P. K. 1986. Cultural aspects of traditional food. In *Traditional foods: Some products and technologies*, CFTRI, 271–279. Central Food Technological Research Institute, Mysore, India.

Miyamoto, M., Y. Seto, D. H. Hao, T. Teshima, Y. B. Sun, T. Kabuki, L. B. Yao, and H. Nakajima. 2005. *Lactobacillus harbinensis* sp. nov., consisted of strains isolated from traditional fermented vegetables *suan cai*, in Harbin, northeastern China and *Lactobacillus perolens* DSM 12745. *Systematic and Applied Microbiology* 28: 688–694.

Mizutani, T., A. Kimizukaa, K. Ruddle, and N. Ishige. 1992. Chemical composition of fermented fish products. *Journal of Food Composition and Analysis* 5: 152–159.

Mohammed, S. I., L. R. Steenson, and A. W. Kirleis. 1991. Isolation and characterization of microorganisms associated with the traditional fermentation of production of Sudanese *kisra*. *Applied and Environmental Microbiology* 57: 2529–2533.

Montel, M. C., F. Masson, and R. Talon. 1998. Bacterial role in flavour development. *Meat Science* 49 (Suppl. 1): S111–S123.

Motwani, M. P., K. C. Jayaram, and K. L. Sehgal. 1962. Fish and fisheries of Brahmaputra River system, Assam: I, Fish fauna with observation on their zoogeographical significance. *Tropical Ecology* 3 (1–2): 17–43.

Mugula, J. K., S. A. M. Ninko, J. A. Narvhus, and T. Sorhaug. 2003. Microbiological and fermentation characteristics of *togwa*, a Tanzanian fermented food. *International Journal of Food Microbiology* 80: 187–199.

Mukherjee, S. K., C. S. Albury, A. G. Pederson, and K. H. Steinkraus. 1965. Role of *Leuconostoc mesenteroides* in leavening the batter of *idli*, a fermented food of India. *Applied Microbiology* 13 (2): 227–231.

Mulyowidarso, R. K., G. H. Fleet, and K. A. Buckle. 1989. The microbial ecology of soybean soaking for *tempe* production. *International Journal of Food Microbiology* 8: 35–46.

Mundt, J. O. 1986. Enterococci. In *Bergy's manual of systematic bacteriology*. Vol. 2, ed. P. H. A. Sneath, N. S. Mair, M. E. Sharpe, and J. G. Holt, 1036–1065. Baltimore, Md.: Williams and Wilkins.

Mundt, J. O. and J. L. Hammer. 1968. Lactobacilli on plant. *Applied Microbiology* 16: 1326–1330.

Nakao, S. 1972. Mame no ryori. In *Ryori no kigen*, 115–126. Tokyo: Japan Broadcast Publishing (Japanese).

Nandy, S. N. and P. K. Sanal. 2005. An outlook of agricultural dependency in the IHR. *ENVIS Newsletter: Himalayan Ecology* 2: 4–5.

Nandy, S. N., P. P. Dhyani, and P. K. Sanal. 2006. Resources information database of the Indian Himalaya. *ENVIS Monograph* 3: 1–95.

Nche, P. F. 1995. Innovations in the production of *kenkey*, a traditional fermented maize product of Ghana: Nutritional, physical and safety aspects. Ph.D thesis, Agricultural University, Wageningen, The Netherlands.

Nche, P. F., G. T. Odamtten, M. J. R. Nout, and F. M. Rombouts. 1994. Dry milling and accelerated fermentation of maize for industrial production of *kenkey*, a Ghanaian cereal food. *Journal of Cereal Science* 20: 291–298.

Nigatu, A., S. Ahrné, B. A. Gashe, and G. Molin. 1998. Randomly amplified polymorphic DNA (RAPD) for discrimination of *Pediococcus pentosaceus* and *Ped. acidilactici* and rapid grouping *Pediococcus* isolates. *Letters in Applied Microbiology* 26: 412–416.

Nikkuni, S. 1997. *Natto, kinema* and *thua-nao*: Traditional non-salted fermented soybean foods in Asia. *Farming Japan* 31 (4): 27–36.

Nikkuni, S., T. B. Karki, K. S. Vilku, T. Suzuki, K. Shindoh, C. Suzuki, and N. Okada. 1995. Mineral and amino acid contents of *kinema*, a fermented soybean food prepared in Nepal. *Food Science and Technology International* 1 (2): 107–111.

Nikkuni, S., T. B. Karki, T. Terao, and C. Suzuki. 1996. Microflora of *mana*, a Nepalese rice *koji*. *Journal of Fermentation and Bioengineering* 81 (2): 168–170.

Niku-Paavola, M.-L., Laitila, A., Mattila-Sandholm, T., and A. Haikara. 1999. New types of antimicrobial compounds produced by *Lactobacillus plantarum*. *Journal of Applied Microbiology* 86: 29–35.

Nostro, A., M. A. Cannatelli, G. Crisafi, A. D. Musolino, F. Procopio, and V. Alonzon. 2004. Modifications of hydrophobicity, *in vitro* adherence and cellular aggregation of *Streptococcus mutans* by *Helichrysum italicum* extract. *Letters in Applied Microbiology* 38: 423–427.

Nout, M. J. R. 1991. Ecology of accelerated natural lactic fermentation of sorghum-based infant food formulas. *International Journal of Food Microbiology* 12: 217–224.

Nout, M. J. R. 1994. Fermented foods and food safety. *Food Research International* 27: 291–298.

Nout, M. J. R. 2001. Fermented foods and their production. In *Fermentation and food safety*, ed. M. R. Adams and M. J. R. Nout, 1–38. Gaithersburg, Md.: Aspen Publishers.

Nout, M. J. R. and K. E. Aidoo. 2002. In *Mycota: A comprehensive treatise on fungi as experimental systems and applied research, industrial applications*. Vol. 10, ed. H. D. Osiewacz, 23–47. Berlin: Springer-Verlag.

Nout, M. J. R. and P. O. Ngoddy. 1997. Technological aspects of preparing affordable fermented complementary foods. *Food Control* 8 (5–6): 279–287.

Nout, M. J. R., B. Kok, E. Vela, P. F. Nche, and F. M. Rombouts. 1996. Acceleration of the fermentation of kenkey, an indigenous fermented maize food of Ghana. *Food Research International* 28(6): 599–604.

Nout, M. J. R., D. Bakshi, and P. K. Sarkar. 1998. Microbiological safety of *kinema*, a fermented soyabean food. *Food Control* 9 (6): 357–362.

Nychas, G. J. E. and J. S. Arkoudelos. 1990. Staphylococci: Their role in fermented sausages. *Journal of Applied Bacteriology Symposium Supplement* 19: 167S–188S.

Odunfa, S. A. 1981. Microorganisms associated with fermentation of African locust bean (*Parkia filicoidea*) during iru preparation. *Journal of Plant Foods* 3: 245–250.

Odunfa, S. A. and E. Y. Adewuyi. 1985. Optimization of process conditions for the fermentation of African locust bean (*Parkia biglobosa*): I, Effect of time, temperature and humidity. *Chemische Mikrobiologie Technologie Lebensmittel* 9: 6–10.

O'Flaherty, W. 1975. *Hindu myths: A sourcebook translated from the Sanskrit*. Harmondsworth, U.K.: Penguin.

Oguntoyinbo, F. A., A. I. Sanni, C. M. A. P. Franz, and W. H. Holzapfel. 2007. *In vitro* fermentation studies for selection and evaluation of *Bacillus* strains as starter cultures for the production of *okpehe*, a traditional African fermented condiment. *International Journal of Food Microbiology* 113 (2): 208–218.

Ohta, T. 1986. Natto. In *Legume-based fermented foods*, ed. N. R. Reddy, M. D. Pierson, and D. K. Salunkhe, 85–95. Boca Raton, Fla.: CRC Press.

Olson, N. F. 1996. Yeasts: Ripened semi-soft cheeses. In *Pfizer Cheese Monographs*. Vol. 4. New York: Pfizer.

Olukoya, D. K., P. S. Tichaczek, A. Butsch, R. F. Vogel, and W. P. Hammes. 1993. Characterization of the bacteriocins produced by *Lactobacillus pentosus* DK7 isolated from *ogi* and *Lactobacillus plantarum* DK9 from *fufu*. *Chemische Mikrobiologie Technologie Lebensmittel* 15: 65–69.

O'Malley, L. S. S. 1907. *Darjeeling District Gazetteer*. New Delhi: Gyan Publishing House.

Omar, N. B., H. Abriouel, R. Lucas, M. Martinez-Canemero, J.-P. Guyot, and A. Galvez. 2006. Isolation of bacteriocinogenic *Lactobacillus plantarum* strains from *ben-saalga*, a traditional fermented gruel from Burkina Faso. *International Journal of Food Microbiology* 112: 44–50.

Onyekwere, O. O., I. A. Akinrele, and O. A. Koleoso. 1989. Industrialization of *ogi* fermentation. In *Industrialization of indigenous fermented foods*, ed. K. H. Steinkraus, 329–362. New York: Marcel Dekker.

Ouwehand, A. C. 1998. Antimicrobial components from lactic acid bacteria. In *Lactic acid bacteria: Microbiology and functional aspects*, ed. S. Salminen and A. von Wright, 139–159. 2nd ed. New York: Marcel Dekker.

Oyewole, O. B. 1997. Lactic fermented foods in Africa and their benefits. *Food Control* 5/6: 289–297.

Pal, P. K., S. A. Hossain, and P. K. Sarkar. 1993. An assessment of manufacturing methods and sensory characteristics of market *churpi*. *Journal of Hill Research* 6: 73–76.

Pal, P. K., S. A. Hossain, P. K. Sarkar, and G. R. Patil. 1994. Compositional and sensory characteristics of *kachcha chhurpi*. *Journal of Food Science and Technology* 31: 71–72.

Pal, P. K., S. A. Hossain, and P. K. Sarkar. 1995. The effect of different coagulants on quality of *chhurpi*. *Indian Journal of Dairy Science* 48: 562–565.

Pal, P. K., S. A. Hossain, and P. K. Sarkar. 1996. Optimisation of process parameters in the manufacture of *churpi*. *Journal of Food Science and Technology* 33: 219–223.

Papamanoli, E., P. Kotzekidou, N. Tzanetakis, and E. L. Tzanetaki. 2002. Characterization of Micrococcaceae isolated from dry fermented sausage. *Food Microbiology* 19: 441–449.

Park, K. I., T. I. Mheen, K. H. Lee, C. H. Chang, S. R. Lee, and T. W. Kwon. 1977. Korean *yakju* and *takju*. In *The Proceeding and Symposium on Indigenous Fermented Foods*, November 21–27, GIAMI, Bangkok.

Paula, C. R., W. Gamble, S. T. Laria, and S. A. Reis Filho. 1998. Occurrence of fungi in butter, available for public consumption in Sao Paul City, *Revistade Microbiology* 19 (3): 317–320.

Pederson, C. S. 1979. *Microbiology of food fermentations*. 2nd ed. Westport, Conn.: AVI Publishing.

Pederson, C. S. and M. N. Albury. 1969. The sauerkraut fermentation. *Food Technology* 8: 1–5.

Phithakpol, B. 1987. *Plaa-raa* traditional Thai fermented fish. In *Proceedings of Conference on Foods and their Processing in Asia*, 182–188. NODAI Research Institute, Tokyo University of Agriculture, Tokyo.

Phithakpol, B. 1993. Fish fermentation technology in Thailand. In *Fish fermentation technology*, ed. C. H. Lee, K. H. Steinkraus, and P. J. A. Reilly, 155–166. Tokyo: United Nations University Press.

Phithakpol, B., W. Varanyanond, S. Reungmaneepaitoon, and H. Wood. 1995. *The traditional fermented foods of Thailand*. Kuala Lumpur, Malaysia: ASEAN Food Handling Bureau.

Pichyangkura, S. and S. Kulprecha. 1977. Survey of mycelial moulds in loog-pang from various sources in Thailand. In *The Proceeding and Symposium on Indigenous Fermented Foods*, November 21–27, GIAMI, Bangkok.

Ponce, A. G., M. R. Moreira, C. E., del Velle, and S. I. Roura. 2008. Preliminary characterization of bacteriocin-like substances from lactic acid bacteria isolated from organic leafy vegetables. *LWT-Food Science and Technology* 41 (3): 432–441.

Pourmorad, F., S. J. Hosseinimehr, and N. Shahabimajd. 2006. Antioxidant activity, phenol and flavonoid contents of some selected Iranian medicinal plants. *African Journal of Biotechnology* 5 (11): 1142–1145.

Pradhan, K. 1982. *The Gorkha conquest: The process and consequences of unification of Nepal with particular references to Eastern Nepal*. Calcutta, India: Oxford University Press.

Pradhan, K. C., E. Sharma, G. Pradhan, and A. B. Chettri. 2004. Geography and environment. *Sikkim Study Series*. Vol. 1, Information and Public Relations Department, Government of Sikkim, Gangtok, India.

Prajapati, J. B. and B. M. Nair. 2003. The history of fermented foods. In *Handbook of fermented functional foods*, ed. R. Farnworth, 1–25. New York: CRC Press.

Prakash, O. 1987. Economy and food in ancient India, Part 2: *Food*. Delhi: Bharatiya Vidya Prakashan.

Pravabati, D. and I. I. T. Singh. 1986. Studies on the chemical and nutritional changes of bamboo shoots during fermentation. *Journal of Food Science and Technology* 23: 338–339.

Pretorius, I. S. 2000. Tailoring wine yeast for the new millennium: Novel approaches to the ancient art of winemaking. *Yeast* 16: 675–729.

Puwastien, P., K. Judprasong, E. Kettwan, K. Vasanachitt, Y. Nakngamanong, and L. Bhattacharjee. 1999. Proximate composition of raw and cooked Thai freshwater and marine fish. *Journal of Food Composition and Analysis* 12: 9–16.

Rai, A. 2008. Microbiology of traditional meat products of Sikkim and Kumaun Himalaya. Ph.D. thesis, Food Microbiology Laboratory, Sikkim Government College and Department of Botany, Kumaun University, Nainital, India.

Rai, A. K., R. M. Sharma, and J. P. Tamang. 2005. Food value of common edible plants of Sikkim. *Journal of Hill Research* 18 (2): 99–103.

Rai, A. K., U. Palni, and J. P. Tamang. 2009. Traditional knowledge of the Himalayan people on production of indigenous meat products. *Indian Journal of Traditional Knowledge* 8 (1): 104–109.

Ramakrishnan, C. V. 1979. Studies on Indian fermented foods. *Baroda Journal of Nutrition* 6: 1–54.

Randazzo, C. L., D. Romano, and C. Caggia. 2004. *Lactobacillus casei*, dominant species in naturally fermented Sicilian green olives. *International Journal of Food Microbiology* 90: 9–14.

Rantsiou, K. and L. Cocolin. 2006. New developments in the study of the microbiota of naturally fermented sausages as determined by molecular methods: A review. *International Journal of Food Microbiology* 108 (2): 255–267.

Reddy, N. R., S. K. Sathe, M. D., Pierson, and D. K. Salunkhe. 1981. *Idli*, an Indian fermented food: A review. *Journal of Food Quality* 5: 89–101.

Reddy, O. V. S. and S. C. Basappa. 1996. Direct fermentation of cassava starch to ethanol by mixed cultures of *Endomycopsis fibuligera* and *Zymomonas mobilis*: Synergism and limitations. *Biotechnology Letters* 18 (11): 1315–1318.

Rigzin, T. 1993. *Festivals of Tibet*. Dharamsala, India: Library of Tibetan Works and Archives.

Risley, H. H. 1928. *The Gazetteer of Sikkim*. New Delhi: D. K. Publishing.

Roberts, T. A., A. C. Baird-Parker, and R. B. Tompkin. 1996. Microorganisms in foods. Vol. 5, *Microbilogical specifications of food pathogens*. London: Blackie Academic and Professional.

Rodriguez, M., F. Nunez, J. J. Cordoba, C. Sanabria, E. Bermudez, and M. A. Asensio. 1994. Characterization of *Staphylococcus* spp. and *Micrococcus* spp. isolated from Iberian ham throughout the ripening process. *International Journal of Food Microbiology* 24: 329–335.

Romano, P., A. Capace, and L. Jespersen. 2006. Taxonomic and ecological diversity of food and beverage yeasts. In *The yeast handbook: Yeasts in food and beverages*, ed. A. Querol, and G. H. Fleet, 13–53. Berlin, Heidelberg: Springer-Verlag.

Rosenberg, M. 1984. Bacterial adherence to hydrocarbons: A useful technique for studying cell-surface hydrophobicity. *FEMS Microbiology Letters* 22: 289–295.

Rosenberg, M., D. Gutnick, and E. Rosenberg. 1980. Adherence of bacteria to hydrocarbons: A simple method for measuring cell-surface hydrophobicity. *FEMS Microbiology Letters* 9: 29–33.

Rubia-Soria, A., H. Abriouel, R. Lucas, N. B. Omar, M. Martinez-Caõamero, and A. Gálvez. 2006. Production of antimicrobial substances by bacteria isolated from fermented table olives. *World Journal of Microbiology and Biotechnology* 22 (7): 765–768.

Salminen, S. and A. V. Wright. 1998. *Lactic acid bacteria microbiology and functional aspects*, 2nd ed. Marcel Dekker, New York.

Samal, P. K., R. Fernando, and D. S. Rawat. 2000. Influences of economy and culture in development among mountain tribes of Indian central Himalaya. *International Journal of Sustainable Development and World Ecology* 7: 41–49.

Samant, S. S. and U. Dhar. 1997. Diversity, endemism and economic potential of wild edible plants of Indian Himalaya. *International Journal of Sustainable Development and World Ecology* 4: 179–191.

Samantray, G. T., P. K. Misra, and K. K. Patnaik. 1989. Mineral composition of ragi. *Indian Journal of Nutrition and Dietetics* 26: 113–116.

Samelis, J., S. Roller, and J. Metaxopoulos. 1994. Sakacin B, a bacteriocin produced by *Lactobacillus sake* isolated from Greek dry fermented sausages. *Journal of Applied Bacteriology* 76: 475–486.

Samelis, J., J. Metaxopoulos, M. Vlassi, and A. Pappa. 1998. Stability and safety of traditional Greek salami: A microbiological ecology study. *International Journal of Food Microbiology* 44: 69–82.

Samson, R. A. 1993. The exploitation of moulds in fermented foods. In *Exploitation of microorganisms*, ed. D. G. Jone, 321–341. London: Chapman and Hall.

Sanchez, P. C. 1996. *Puto*: Philippine fermented rice cake. In *Handbook of indigenous fermented food*, ed. K. H. Steinkraus, 167–182. 2nd ed. New York: Marcel Dekker.

Sandine, W. E., P. C. Radich, and P. R. Elliker. 1972. Ecology of the lactic strepto-cocci: A review. *Journal of Milk Food Technology* 35: 176–184.

Santos, E. M., C. González-Fernández, I. Jaime, and J. Rovira. 1998. Comparative study of lactic acid bacteria house flora isolated in different varieties of *chorizo. International Journal of Food Microbiology* 39 (1–2): 123–128.

Saono, S., L. Gandjar, T. Basuki, and H. Karsono. 1974. Mycoflora of *ragi* and some other traditional fermented foods of Indonesia. *Annales Bogorienses* 5 (4): 187–204.

Sarangthem, K. and T. N. Singh. 2003. Microbial bioconversion of metabolites from fermented succulent bamboo shoots into phytosterols. *Current Science* 84 (12): 1544–1547.

Sarkar, P. K. and J. P. Tamang. 1994. The influence of process variables and inoc-ulum composition on the sensory quality of kinema. *Food Microbiology* 11: 317–325.

Sarkar, P. K. and J. P. Tamang. 1995. Changes in the microbial profile and proximate composition during natural and controlled fermentations of soybeans to pro-duce *kinema. Food Microbiology* 12: 317–325.

Sarkar, P. K., J. P. Tamang, P. E. Cook, and J. D. Owens. 1994. *Kinema*: A tradi-tional soybean fermented food: Proximate composition and microflora. *Food Microbiology* 11: 47–55.

Sarkar, P. K., L. J. Jones, W. Gore, and G. S. Craven. 1996. Changes in soya bean lipid profiles during *kinema* production. *Journal of Science of Food and Agriculture* 71: 321–328.

Sarkar, P. K., L. J. Jones, G. S. Craven, S. M. Somerset, and C. Palmer. 1997a. Amino acid profiles of *kinema*, a soybean-fermented food. *Food Chemistry* 59 (1): 69–75.

Sarkar, P. K., L. J. Jones, G. S. Craven, and S. M. Somerset. 1997b. Oligosaccharides profile of soybeans during kinema production. *Letters in Applied Microbiology* 24: 337–339.

Sarkar, P. K., E. Morrison, U. Tingii, S. M. Somerset, and G. S. Craven. 1998. B-group vitamin and mineral contents of soybeans during *kinema* production. *Journal of Science of Food and Agriculture* 78: 498–502.

Sarkar, P. K., B. Hasenack, and M. J. R. Nout. 2002. Diversity and functionality of *Bacillus* and related genera isolated from spontaneously fermented soybeans (Indian *kinema*) and locust beans (African *soumbala). International Journal of Food Microbiology* 77: 175–186.

Savitri, T. and C. Bhalla. 2007. Traditional foods and beverages of Himachal Pradesh. *Indian Journal of Traditional Knowledge* 6 (1): 17–24.

Schillinger, U. and F. K. Lücke. 1987. Identification of lactobacilli from meat and meat products. *Food Microbiology* 4: 199–208.

Schillinger, U. and F. K. Lücke. 1989. Antibacterial activity of *Lactobacillus sake* iso-lated from meat. *Applied and Environmental Microbiology* 55 (8): 1901–1906.

Schillinger, U., W. Holzapfel, and O. Kandler. 1989. Nucleic acid hybridization studies on *Leuconostoc* and heterofermentative lactobacilli and description of *Leuconostoc amelibiosum* sp. nov. *Systematic and Applied Microbiology* 12: 48–55.

Schillinger, U., N. M. K. Yousif, L. Sesar, and C. M. A. P. Franz. 2003. Use of group-specific and RAPD-PCR analyses for rapid differentiation of *Lactobacillus* strains from probiotic yogurts. *Current Microbiology* 47: 453–456.

Schillinger, U., C. Guigas, J. P. Tamang, C. M. A. P. Franz, M. Gores, S. Hucker, D. Vogel, and W. H. Holzapfel. 2004. Identification and determination of some functional properties of lactic acid bacteria isolated from traditionally fermented vegetables of Sikkim. In *Proceeding Abstract of the 19th International ICFMH Symposium Food Micro 2004 on "New tools for improving microbial food safety and quality: Biotechnology and molecular biology approaches,"* Portorož, Slovenia, Sep. 12–16, 2004, p. 383.

Selvakumar, P., L. Ashakumary, and A. Pandey. 1996. Microbial synthesis of starch saccharifying enzyme in solid cultures. *Journal of Scientific and Industrial Research* 55: 443–449.

Shah, N. P. 1994. *Lactobacillus acidophilus* and lactose intolerance: A review. *ASEAN Food Journal* 9: 47–54.

Shah, N. P. 2001. Functional foods from probiotics and prebiotics. *Food Technology* 55 (11): 46–53.

Shah, N. P. 2004. Probiotics and prebiotics. *Agro Food Industry Hi-Tech* 15 (1): 13–16.

Shah, N. P. 2005. Fermented functional foods: An overview. In *The Proceeding of the Second International Conference on Fermented Foods, Health Status and Social Well-being*, organized by Swedish South Asian Network on Fermented Foods and Anand Agricultural University, Anand, Gujarat, India, Dec 17–18, 2005, 1–6.

Shah, N. P. 2007. Functional cultures and health benefits. *International Dairy Journal* 17: 1262–1277.

Shah, N. P. and P. Jelen. 1990. Survival of lactic acid bacteria and their lactases under acidic conditions. *Journal of Food Science* 55: 506–509.

Sharma, D. K., K. Ghosh, M. Raquib, and M. Bhattacharya. 2006. Yak products' profile: An overview. *Journal of Food Science and Technology* 43: 447.

Shon, M.-Y., J. Lee, J.-H. Choi, S. Y. Choi, S.-H. Nam, K.-I. Seo, S.-W. Lee, N.-J. Sung, and S.-K. Park. 2007. Antioxidant and free radical scavenging activity of methanol extract of *chungkukjang*. *Journal of Food Composition and Analysis* 20: 113–118.

Shrestha, A. K. and A. Noomhorn. 2001. Composition and functional properties of fermented soybean flour (*kinema*). *Journal of Food Science Technology* 38 (5): 467–470.

Shrestha, A. K. and A. Noomhorn. 2002. Comparison of physicochemical properties of biscuits supplemented with soya and *kinema* flours. *International Journal of Food Science Technology* 37 (4): 361–368.

Shrestha, H. and E. R. Rati. 2003. Defined microbial starter formulation for the production of *poko*, a traditional fermented food product of Nepal. *Food Biotechnology* 17 (1): 15–25.

Shrestha, H., K. Nand, and E. R. Rati. 2002. Microbiological profile of *murcha* starters and physicochemical characteristics of *poko*, a rice based traditional food products of Nepal. *Food Biotechnology* 16: 1–15.

Silla-Santos, M. H. 2001. Toxic nitrogen compounds produced during processing: Biogenic amines, ethyl carbamides, nitrosamines. In *Fermentation and food safety*, ed. M. R. Adams and M. J. R. Nout, 119–140. Gaithersburg, Md.: Aspen Publishers.

Simoncini, N., D. Rotelli, R. Virgili, and S. Quuintavalla. 2007. Dynamics and characterization of yeasts during ripening of typical Italian dry-cured ham. *Food Microbiology* 24 (6): 577–584.

Simpson, W. J. and H. Taguchi. 1995. The genus *Pediococcus*, with notes on the genera *Tetragenococcus* and *Aerococcus*. In *The genera of lactic acid bacteria*, ed. B. J. Wood and W. H. Holzapfel, 125–172. London: Blackie Academic and Professional.

Singh, K. A. 2002. Boon of bamboo resources in North East India. In *Resource management perspective of Arunachal Pradesh*, ed. K. A. Singh, 69–112. Shillong, India: ICAR Research Complex for NEH Region.

Singh, N. I. and A. Umabati Devi. 1995. Fermentation prospects of two phylloplane bacteria in traditional *hawaijar* made from boiled soybean (*Glycine max* L.). *Journal of Food Science and Technology* 32 (3): 219–220.

Singh, P. K. and K. I. Singh. 2006. Traditional alcoholic beverage, *yu* of Meitei communities of Manipur. *Indian Journal of Traditional Knowledge* 5 (2): 184–190.

Singh, R. L. 1991. *India: A regional geography.* Varanasi: National Geographical Society of India.

Smith, J. L. and S. A. Palumbo. 1983. Use of starter cultures in meats. *Journal of Food Protection* 46: 997–1006.

Soedarsono, J. 1972. Some notes on *ragi tape*, an inoculum for *tapé* fermentation. *Majalah Ilmu Pertanian* 1: 235–241.

Song, D. J. and Y. H. Park. 1992. Effect of lactic acid bacteria on the growth of yeast from *mool kimchi*. *Korean Journal of Applied Microbiology and Biotechnology* 20 (2): 219.

Soni, S. K. and D. K. Sandhu. 1989a. Nutritional improvement of Indian dosa batter by yeast enrichment and black gram replacement. *Journal of Fermentation and Bioengineering* 68 (1): 1–4.

Soni, S. K. and D. K. Sandhu. 1989b. Fermentation of idli: Effects of changes in raw materials and physicochemical conditions. *Journal of Cereal Science* 10: 227–238.

Soni, S. K. and D. K. Sandhu. 1990a. Biochemical and nutritional changes associated with Indian Punjab *wari* fermentation. *Journal of Food Science and Technology* 27: 82–85.

Soni, S. K. and D. K. Sandhu. 1990b. Indian fermented foods: Microbiological and biochemical aspects. *Indian Journal of Microbiology* 30: 130–157.

Soni, S. K. and D. K. Sandhu. 1991. Role of yeast domination in Indian *idli* batter fermentation. *World Journal of Microbiology and Biotechnology* 7: 505–507.

Soni, S. K., D. K. Sandhu, and K. S. Vilkhu. 1985. Studies on *dosa*: An indigenous Indian fermented food: Some biochemical changes accompanying fermentation. *Food Microbiology* 2: 175–181.

Soni, S. K., D. K. Sandhu, K. S. Vilkhu, and N. Kamra. 1986. Microbiological studies on dosa fermentation. *Food Microbiology* 3: 45–53.

Spinler, J. K., M. Taweechotipatr, C. L. Rognerud, C.N. Ou, S. Tumwasorn, and J. Vorsalovic. 2008. Human derived probiotic *Lactobacillus reuteri* demonstrate antimicrobial activities targeting enteric bacterial pathogens. *Anaerobe* 14 (3): 166–171.

Steinkraus, K. H. 1983. Lactic acid fermentation in the production of foods from vegetables, cereals and legumes. *Antonie van Leeuwenhoek* 49: 337–348.

Steinkraus, K. H. 1994. Nutritional significance of fermented foods. *Food Research International* 27: 259–267.

Steinkraus, K. H. 1996. *Handbook of indigenous fermented food.* 2nd ed. New York: Marcel Dekker.

Steinkraus, K. H. 1997. Classification of fermented foods: Worldwide review of household fermentation techniques. *Food Control* 8 (5–6): 331–317.

Steinkraus, K. H., A. G. van Veer, and D. B. Thiebeau. 1967. Studies on *idli*: An Indian fermented black gram-rice food. *Food Technology* 21 (6): 110–113.

Stiles, M. E. and W. H. Holzapfel. 1997. Lactic acid bacteria of foods and their current taxonomy. *International Journal of Food Microbiology* 36: 1–29.

Stratton, J. E., R. W. Hutkins, and S. L. Taylor. 1991. Biogenic amines in cheese and other fermented foods: A review. *Journal of Food Protection* 54: 460–470.

Subba, C. 1999. Nepal ko samajik samrachana ra sanskritik sambardhan ko sawal. *Nationalities* (National Committee for Development of Nationalities) 1 (1–2): 64 (Nepali).

Subba, J. R. 1999. *The Limboos of the Eastern Himalayas with special reference to Sikkim*. Gangtok, India: Sukhim Yakthung Mundhum Saplopa.

Subba, J. R. 2008. *History, culture and customs of Sikkim*. New Delhi: Gyan Publishing House.

Sukhumavasi, J., K. Kato, and T. Harada. 1975. Glucoamylase of a strain of *Endomycopsis fibuligera* isolated from mould bran (*look pang*) of Thailand. *Journal of Fermentation Technology* 53 (8): 559–565.

Sundriyal, M. and L. K. Rai. 1996. Wild edible plants of Sikkim Himalaya. *Journal of Hill Research* 9 (2): 267–278.

Sundriyal, M. and R. C. Sundriyal. 2004. Wild edible plants of the Sikkim Himalaya: Nutritive values of selected species. *Economic Botany* 58 (2): 286–299.

Suprianto, O. R., T. Koga, and S. Ueda. 1989. Liquefaction of glutinous rice and aroma formation in *tapé* preparation by *ragi*. *Journal of Fermentation and Bioengineering* 64 (4): 249–252.

Suzzi, G. and F. Gardini. 2003. Biogenic amines in dry fermented sausages: A review. *International Journal Food Microbiology* 88: 41–45.

Svanberg, U., W. Lorri, and A.-S. Sandberg. 1993. Lactic fermentation of non-tannin and high-tannin cereals: Effects on in vitro estimation of iron availability and phytate hydrolysis. *Journal of Food Science* 58: 408–412.

Tagg, J. R. 1992. Bacteriocins of Gram-positive bacteria: An opinion regarding their nature, nomenclature and numbers. In *Bacteriocins, microcins and lantibiotics*. Vol. H65, ed. R. James, C. Lazdunski, and F. Pattus, 33–35. Berlin: NATO ASI Series, Springer.

Tagg, J. R., A. S. Dajani, and L. W. Wannamaker. 1976. Bacteriocins of Gram-positive bacteria. *Bacteriological Reviews* 40: 722–756.

Tamang, B. 2006. Role of lactic acid bacteria in fermentation and biopreservation of traditional vegetable products. Ph.D. thesis, Food Microbiology Laboratory, Sikkim Government College, North Bengal University.

Tamang, B. and J. P. Tamang. 2007. Role of lactic acid bacteria and their functional properties in *goyang*, a fermented leafy vegetable product of the Sherpas. *Journal of Hill Research* 20 (20): 53–61.

Tamang, B. and J. P. Tamang. 2009a. Traditional knowledge of biopreservation of perishable vegetable and bamboo shoots in Northeast India as food resources. *Indian Journal of Traditional Knowledge* 8 (1): 89–95.

Tamang, B. and J. P. Tamang. 2009b. Lactic acid bacteria isolated from indigenous fermented bamboo products of Arunachal Pradesh in India and their functionality. *Food Biotechnology* 23: 133–147.

Tamang, B. and J. P. Tamang. 2009c. *In situ* fermentation dynamics during production of *gundruk* and *khalpi*, ethnic fermented vegetables products of the Himalayas. *Indian Journal of Microbiology* (in press).

Tamang, B., J. P. Tamang, U. Schillinger, C. M. A. P. Franz, M. Gores, and W. H. Holzapfel. 2008. Phenotypic and genotypic identification of lactic acid bacteria isolated from ethnic fermented tender bamboo shoots of North East India. *International Journal of Food Microbiology* 121: 35–40.

Tamang, J. P. 1982. A brief historical account of Darjeeling. Pines and Camellias, Darjeeling Government College, Darjeeling, 8–15.

Tamang, J. P. 1992. Studies on the microflora of some traditional fermented foods of the Darjeeling hills and Sikkim. Ph.D. thesis, North Bengal University, Darjeeling.

Tamang, J. P. 1994. Kinema: A non-salted soybean fermented food of Darjeeling hills and Sikkim. In *Proceeding Abstract of the Third Asian Symposium on Non-Salted Soybean Fermentation*, Akita, Japan, June 4–6, 1999, p. 14.

Tamang, J. P. 1995. Study of traditional fermented foods production in the Darjeeling hills and Sikkim, with emphasis on kinema. Post-doctorate dissertation (food and nutrition), National Food Research Institute, Tsukuba, under the United Nations University fellowship program, Japan.

Tamang, J. P. 1996. Fermented soybean products in India. In *The Proceedings of the 2nd International Soybean Processing and Utilization Conference*, ed. A. Buchanan. Bangkok, Thailand, January 8–13, 1996, pp. 189–193.

Tamang, J. P. 1997. Jatiya Khathyaharu: *Kinema, Gundruk* and *Sinki*. In *Nirman*, ed. P. Chamling and G. Gaule, 189–193. Namchi, Nepali: Nirman Prakhashan.

Tamang, J. P. 1998a. Role of microorganisms in traditional fermented foods. *Indian Food Industry* 17 (3): 162–167.

Tamang, J. P. 1998b. Upgradation of kinema production for sustainable development of protein-rich soybean food in the Himalayan regions of the Darjeeling hills and Sikkim. Final report of the UNU-Kirin Brewery Pvt. Ltd., Japan, Follow-up Research Programme, Gangtok, India.

Tamang, J. P. 1999. Development of pulverised starter for *kinema* production. *Journal of Food Science and Technology* 36 (5): 475–478.

Tamang, J. P. 2000a. Case study on socio-economical prospective of *kinema*, a traditional fermented soybean food. In *Proceedings of the 1997 International Conference on Traditional Foods*, CFTRI, 180-185. March 6–8, 1997, Central Food Technological Research Institute, Mysore.

Tamang, J. P. 2000b. Traditional fermented foods and beverages of the Sikkim Himalayas in India: Indigenous process and product characterization. In *Proceedings of the International Conference on Traditional Foods*, CFTRI, 99-116. March 6–8, 1997, Central Food Technological Research Institute, Mysore.

Tamang, J. P. 2000c. Microbial diversity associated with natural fermentation of *kinema*. In *Proceedings of the Third International Soybean Processing and Utilization Conference*, ed. K. Saio, 713–717. Tsukuba, Japan, October 15–20, 2000.

Tamang, J. P. 2001a. Food culture in the Eastern Himalayas. *Journal of Himalayan Research and Cultural Foundation* 5 (3–4): 107–118.

Tamang, J. P. 2001b. Kinema. *Food Culture* 3: 11–14.

Tamang, J. P. 2002. Lesser-known ethnic fermented soybean foods of the Eastern Himalayas. In *Proceeding Abstract of the China & International Soybean Conference, American Oil Chemists' Society and Chinese Cereals and Oils Association*, Beijing, China, pp. 232–233.

Tamang, J. P. 2003. Native microorganisms in fermentation of kinema. *Indian Journal of Microbiology* 43(2): 127–130.

Tamang, J. P. 2005a. *Food culture of Sikkim*. Sikkim Study Series vol. 4. Information and Public Relations Department, Government of Sikkim, Gangtok, India.

Tamang, J. P. 2005b. Ethnic fermented foods of the Eastern Himalayas. In *Proceedings of the Second International Conference on "Fermented Foods, Health Status and Social Well-being,"* Dec. 17–18, 2005, pp. 19–26. Organized by Swedish South Asian Network on Fermented Foods and Anand Agricultural University, Anand, Gujarat.

Tamang, J. P. 2005c. *Carrying capacity study of Teesta Basin in Sikkim.* Vol. 7, *Food resources: Edible wild plants and ethnic fermented foods.* Final project report in collaboration with South Campus-Delhi University, New Delhi.

Tamang, J. P. 2007a. Fermented foods for human life. In *Microbes for human life,* ed. A. K. Chauhan, A. Verma, and H. Kharakwal, 73–87. New Delhi, India: I.K. International Publishing.

Tamang, J. P. 2007b. History of scientific developments in Darjeeling Hills. In *Discursive Hills studies in history, polity and economy,* ed. P. J. Victor, P. Pradhan, D. Lama, and A. Das, 237–241. Darjeeling, India: St. Joseph's College.

Tamang, J. P. 2009. Food and identity: a study among the Nepalis of Sikkim and Darjeeling. In *Indian Nepalis,* ed. T. B. Subba, A. C. Sinha, G. S. Nepal, and D. R. Nepal, 297–310. New Delhi, India: Concept Publishing.

Tamang, J. P. and P. K. Sarkar. 1993. *Sinki*: A traditional lactic acid fermented radish taproot product. *Journal of General and Applied Microbiology* 39: 395–408.

Tamang, J. P. and P. K. Sarkar. 1995. Microbiology of *murcha*: An amylolytic fermentation starter. *Microbios* 81: 115–122.

Tamang, J. P. and S. Nikkuni. 1996. Selection of starter culture for production of *kinema*, fermented soybean food of the Himalaya. *World Journal of Microbiology and Biotechnology* 12 (6): 629–635.

Tamang, J. P. and P. K. Sarkar. 1996. Microbiology of *mesu*, a traditional fermented bamboo shoot product. *International Journal of Food Microbiology* 29: 49–58.

Tamang, J. P. and S. Nikkuni. 1998. Effect of temperatures during pure culture fermentation of *kinema*. *World Journal of Microbiology and Biotechnology* 14 (6): 847–850.

Tamang, J. P. and N. Tamang. 1998. Traditional food recipes of the Sikkim Himalayas. Report, Sikkim Biodiversity and Eco-tourism Project, GBPIHED, Tadong.

Tamang, J. P. and W. H. Holzapfel. 1999. Microfloras of fermented foods. In *Encyclopedia of food microbiology.* Vol. 2, ed. R. K. Robinson, C. A. Bhatt, and P. D. Patel, 249–252. London: Academic Press.

Tamang, J. P. and W. H. Holzapfel. 2004. Role of lactic acid bacteria in fermentation, safety and quality of traditional vegetable products in the Sikkim Himalayas. Final Project Report, Volkswagen Foundation, Karlsruhe, Germany.

Tamang, J. P. and S. Thapa. 2006. Fermentation dynamics during production of *bhaati jaanr*, a traditional fermented rice beverage of the Eastern Himalayas. *Food Biotechnology* 20 (3): 251–261.

Tamang, J. P. and G. H. Fleet. 2009. Yeasts diversity in fermented foods and beverages. In *Yeasts biotechnology: Diversity and applications,* ed. T. Satyanarayana and G. Kunze. New York: Springer (in press).

Tamang, J. P., P. K. Sarkar, and C. W. Hesseltine. 1988. Traditional fermented foods and beverages of Darjeeling and Sikkim: A review. *Journal of Science of Food and Agriculture* 44: 375–385.

Tamang, J. P., S. Thapa, N. Tamang, and B. Rai. 1996. Indigenous fermented food beverages of Darjeeling hills and Sikkim: Process and product characterization. *Journal of Hill Research* 9 (2): 401–411.

Tamang, J. P., S. Dewan, S. Thapa, N. A. Olasupo, U. Schillinger, and W. H. Holzapfel. 2000. Identification and enzymatic profiles of predominant lactic acid bacteria isolated from soft-variety *chhurpi*, a traditional cheese typical of the Sikkim Himalayas. *Food Biotechnology* 14 (1–2): 99–112.

Tamang, J. P., S. Thapa, S. Dewan, J. Yasuka, R. Fudou, and S. Yamanaka. 2002. Phylogenetic analysis of *Bacillus* strains isolated from fermented soybean foods of Asia: *Kinema, chungkokjang* and *natto*. *Journal of Hill Research* 15 (2): 56–62.

Tamang, J. P., S. Dewan, and W. H. Holzapfel. 2004. Technological properties of predominant lactic acid bacteria isolated from indigenous fermented milk products of Sikkim in India. In *Proceeding Abstract of the 19th International ICFMH Symposium Food Micro 2004 on "New Tools for Improving Microbial Food Safety and Quality: Biotechnology and Molecular Biology Approaches,"* organized by University of Ljubljana, Slovenian Microbiological Society, and International Committee for Food Microbiology and Hygiene, Portorož, Slovenia, Sep. 12–16, 2004, p. 300.

Tamang, J. P., B. Tamang, U. Schillinger, C. M. A. P. Franz, M. Gores, and W. H. Holzapfel. 2005. Identification of predominant lactic acid bacteria isolated from traditional fermented vegetable products of the Eastern Himalayas. *International Journal of Food Microbiology* 105 (3): 347–356.

Tamang, J. P., S. Dewan, B. Tamang, A. Rai, U. Schillinger, and W. H. Holzapfel. 2007a. Lactic acid bacteria in *hamei* and *marcha* of North East India. *Indian Journal of Microbiology* 47 (2): 119–125.

Tamang, J. P., N. Thapa, B. Rai, S. Thapa, H. Yonzan, S. Dewan, B. Tamang, R. M. Sharma, A. K. Rai, R. Chettri, B. Mukhopadhyay, and B. Pal. 2007b. Food consumption in Sikkim with special reference to traditional fermented foods and beverages: A micro-level survey. *Journal of Hill Research*, Suppl. issue 20 (1): 1–37.

Tamang, J. P., R. Chettri, and R. M Sharma. 2009. Indigenous knowledge of Northeast women on production of ethnic fermented soybean foods. *Indian Journal of Traditional Knowledge* 8 (1): 122–126.

Tanaka, T. 2008a. Pepok. In *Advanced science on natto*, ed. K. Kiuchi, T. Nagai, and K. Kimura, 218–221. Tokyo: Kenpakusha (Japanese).

Tanaka, T. 2008b. Sieng. In *Advanced science on natto*, ed. K. Kiuchi, T. Nagai, and K. Kimura, 221–224. Tokyo: Kenpakusha (Japanese).

Tanaka, T. and N. Okazaki. 1982. Growth of mould on uncooked grain. *Hakkokogaku* 60: 11 17.

Tanasupawat, S., S. Okada, K. Suzuki, M. Kozaki, and K. Komagata. 1992. Identification of *Enterococcus hirae, E. faecalis, E. faecium*, and *E. casseliflavus* strains from fermented foods. *Bulletin of the Japan Federation for Culture Collections* 8 (2): 86–94.

Tanasupawat, S., S. Okada, K. Suzuki, M. Kozaki, and K. Komagata. 1993. Lactic acid bacteria, particularly heterofermentative lactobacilli, found in fermented foods of Thailand. *Bulletin of the Japan Federation for Culture Collections* 9: 65–78.

Tanimura, W., P. C. Sanchez, and M. Kozaki. 1978. The fermented foods in the Philippines; Part 2: *Basi* (sugarcane wine). *Journal of Agricultural Society (Japan)* 22: 118–133.

Taylor, S. L. 1986. Histamine food poisoning: Toxicology and clinical aspects. *CRC Critical Review in Toxicology* 17: 91–128.

Taylor, S. L., M. Leatherwood, and E. R. Lieber. 1978. Histamine in sauerkraut. *Journal of Food Science* 43: 1030–1032.

Tee, E.-S., M. C. Dop, and P. Winichagoon. 2004. Proceedings of the workshop on food-consumption surveys in developing countries: Future challenges. *Food and Nutrition Bulletin* 25 (4): 407–414.

ten Brink, B., C. Damink, H. J. Joosten, and J. Huis in't Veld. 1990. Occurrence and formation of biologically active amines in foods. *International Journal of Food Microbiology* 11: 73–84.

Teramoto, Y., S. Yoshida, and S. Ueda. 2002. Characteristics of a rice beer (*zutho*) and a yeast isolated from the fermented product in Nagaland, India. *International Journal of Food Microbiology* 18 (9): 813–816.

Teuber, M., A. Geis, and H. Neve. 1991. The genus *Lactococcus*. In *The prokaryotes*. Vol. 2, ed. A. Balows, H. G. Truper, M. Dworkin, W. Harder, and K.-H. Schleifer, 1482–1501. 2nd ed. New York: Springer-Verlag.

Thakur, N., C. Savitri, and T. Bhalla. 2004. Characterization of some traditional fermented foods and beverages of Himachal Pradesh. *Indian Journal of Traditional Knowledge* 3 (3): 325–335.

Thapa, N. 2002. Studies on microbial diversity associated with some fish products of the Eastern Himalayas. Ph.D. thesis, North Bengal University, Darjeeling.

Thapa, N. and J. Pal. 2007. Proximate composition of traditionally processed fish products of the Eastern Himalayas. *Journal of Hill Research* 20 (2): 75–77.

Thapa, N., J. Pal, and J. P. Tamang. 2004. Microbial diversity in *ngari, hentak* and *tungtap*, fermented fish products of Northeast India. *World Journal of Microbiology and Biotechnology* 20 (6): 599–607.

Thapa, N., J. Pal, and J. P. Tamang. 2006. Phenotypic identification and technological properties of lactic acid bacteria isolated from traditionally processed fish products of the Eastern Himalayas. *International Journal of Food Microbiology* 107 (1): 33–38.

Thapa, N., J. Pal, and J. P. Tamang. 2007. Microbiological profile of dried fish products of Assam. *Indian Journal of Fisheries* 54 (1): 121–125.

Thapa, S. 2001. Microbiological and biochemical studies of indigenous fermented cereal-based beverages of the Sikkim Himalayas. Ph.D. thesis, Food Microbiology Laboratory, Sikkim Government College (under North Bengal University), p. 190.

Thapa, S. and J. P. Tamang. 2004. Product characterization of *kodo ko jaanr*: Fermented finger millet beverage of the Himalayas. *Food Microbiology* 21: 617–622.

Thapa, S. and J. P. Tamang. 2006. Microbiological and physicochemical changes during fermentation of *kodo ko jaanr*, a traditional alcoholic beverage of the Darjeeling hills and Sikkim. *Indian Journal of Microbiology* 46 (4): 333–341.

Tichaczek, P. S., J. Nissen-Meyer, I. F. Nes, R. F. Vogel, and W. P. Hammes. 1992. Characterization of the bacteriocin curvacin A from *Lactobacillus curvatus* LTH174 and sakacin P from *Lactobacillus sake* LTH673. *Systematic and Applied Microbiology* 15: 460–468.

Tolonen, M., S. Rajaniemi, J.-M. Pihlava, T. Johansson, P. E. J. Saris, and E.-L. Ryhänen. 2004. Formation of nisin, plant-derived biomolecules and antimicrobial activity in starter culture fermentations of sauerkraut. *Food Microbiology* 21: 167–179.

Tongananta, Q. and C. A. Orillo. 1996. Studies on the Philippine fermented rice cake puto. In *Handbook of indigenous fermented food*, ed. K. H. Steinkraus, 167–182. 2nd ed. New York: Marcel Dekker.

Tou, E. H., J. P. Guyot, C. Mouquet-River, I. Rochette, E. Counil, A. S. Traore, and S. Treche. 2006. Study through surveys and fermentation kinetics of the traditional processing of pearl millet *(Pennisetum glaucum)* into *ben-saalga*, a fermented gruel from Burkina Faso. *International Journal of Food Microbiology* 106: 52–60.

Tou, E. H., C. Mouquet-River, I. Rochette, A. S. Traore, S. Treche, and J. P. Guyot. 2007. Effect of different process combinations on the fermentation kinetics, microflora and energy density of *ben-saalga*, a fermented gruel from Burkina Faso. *Food Chemistry* 100: 935–943.

Tsai, Y.-H., H.-F. Kung, S.-C. Chang, T.-M. Lee, and C.-I. Wei. 2007. Histamine formation by histamine-forming bacteria in *douchi*, a Chinese traditional soybean product. *Food Chemistry* 103 (2007) 1305–1311.

Tsai, Y. W. and S. C. Ingham. 1997. Survival of *E. coli* O157:H7 and *Salmonella* spp. in acidic condiments. *Journal of Food Protection* 60: 751–755.

Tsuyoshi, N., R. Fudou, S. Yamanaka, M. Kozaki, N. Tamang, S. Thapa, and J. P. Tamang. 2005. Identification of yeast strains isolated from *marcha* in Sikkim, a microbial starter for amylolytic fermentation. *International Journal of Food Microbiology* 99 (2): 135–146.

Tyn, M. T. 1993. Trends of fermented fish technology in Burma. In *Fish fermentation technology*, ed. C. H. Lee, K. H. Steinkraus, and P. J. A. Reilly, 129–153. Tokyo: United Nations University Press.

Uchimura, T., Y. Kojima, and M. Kozaki. 1990. Studies on the main saccharifying microorganism in the Chinese starter of Bhutan, *chang poo*. *Journal of Brewing Society of Japan* 85 (12): 881–887.

Uchimura, T., S. Okada, and M. Kozaki. 1991. Identification of lactic acid bacteria isolated from Indonesian Chinese starter, *ragi*; Part 4: Microorganisms in Chinese starters from Asia. *Journal of Brewing Society of Japan* 86 (1): 55–61.

Ueda, S. 1989. Utilization of soybean as natto, a traditional Japanese food. In Bacillus subtilis: *Molecular biology and industrial application*, ed. B. Maruo and H. Yoshikawa, 143–161. Tokyo: Elsevier.

Ueda, S. and S. Kano. 1975. Multiple forms of glucoamylase of *Rhizopus* species. *Die Stärke* 27 (4): 123–128.

Uhlman, L., U. Schillinger, J. R. Rupnow, and W. H. Holzapfel. 1992. Identification and characterization of two bacteriocin-producing strains of *Lactococcus lactis* isolated from vegetables. *International Journal of Food Microbiology* 16: 141–151.

Umeta, M., C. E. West, and H. Fufa. 2005. Content of zinc, iron, calcium and their absorption inhibitors in foods commonly consumed in Ethiopia. *Journal of Food Composition and Analysis* 18: 803–817.

UNWTO. 2008. *UNWTO World Tourism Barometer* 5, no. 2.

Urushibata, Y., S. Tokuyama, and Y. Tahara. 2002. Characterization of the *Bacillus subtilis ywsC* gene involved in λ–polyglutamic acid production. *Journal of Bacteriology* 184 (2): 337–343.

Vachanavinich, K., W. J. Kim, and Y. I. Park. 1994. Microbial study on *krachae*, Thai rice wine. In *Lactic acid fermentation of non-alcoholic dairy food and beverages*, ed. C. H. Lee, J. Adler-Nissen, and G. Bärwald, 233–246. Seoul: Ham Lim Won.

Valdez, G. F. de, G. S. de Giori, M. Garro, F. Mozzi, and G. Oliver. 1990. Lactic acid bacteria from naturally fermented vegetables. *Microbiologie-Aliments-Nutrition* 8: 175–179.

Van Loosdrecht, M. C. M., J. Lyklema, W. Norde, G. Schraa, and A. J. B. Zehnder. 1987. The role of bacterial cell wall hydrophobicity in adhesion. *Applied and Environmental Microbiology* 53 (8): 1893–1897.

Vaughn, R. H. 1985. The microbiology of vegetable fermentations. In *Microbiology of fermented foods*. Vol. 1, ed. B. J. B. Wood, 49–109. London: Elsevier Applied Science.

Vaughn, R. H., K. E. Stevenson, B. A. Dave, and H. C. Park. 1972. Fermenting yeasts associated with softening and gas-pocket formation in olives. *Applied Microbiology* 23: 316–320.

Villar, I., M. C. Garcia-Fontan, B. Prieto, M. E. Tornadizo, and J. Carballo. 2000. A survey on the microbiological changes during the manufacture of dry-cured lacon, a Spanish traditional meat product. *Journal of Applied Microbiology* 89: 1018–1026.

Vinderola, C. G., M. Medici. and G. Perdigón. 2004. Relationship between interaction sites in the gut, hydrophobicity, mucosal immunomodulating capacities and cell wall protein profiles in indigenous and exogenous bacteria. *Journal of Applied Microbiology* 96: 230–243.

Wang, H. L. and C. W. Hesseltine. 1970. Sufu and lao-chao. *Journal of Agricultural and Food Chemistry* 18: 572–575.

Wang, H. L. and C. W. Hesseltine. 1981. Use of microbial cultures: Legume and cereal products. *Food Technology* 35: 79–83.

Wang, H. L., E. W. Swain, L. L. Wallen, and C. W. Hesseltine. 1975. Free fatty acids identified as antitryptic factor in soybeans fermented by *Rhizopus oligosporos*. *Journal of Nutrition* 105: 1351–1355.

Wang, H. L., E. W. Swain, C. W. Hesseltine, and H. D. Heath. 1979. Hydration of hole soybeans affects solid losses and cooking quality. *Journal of Food Science* 44: 1510–1513.

Wang, L.-J., D. Li, L. Zou, X. D. Chen, Y.-Q. Cheng, K. Yamaki, and L.-T. Li. 2007. Antioxidative activity of douchi (a Chinese traditional salt-fermented soybean food) extracts during its processing. *International Journal of Food Properties* 10: 1–12.

Wardlaw, G. M., P. M. Insel, and M. F. Seyler. 1994. *Contemporary Nutrition Issues and Insights*. 2nd ed. London: Masby.

Watanaputi, S. P., R. Chanyavongse, S. Tubplean, S. Tanasuphavatana, and S. Srimahasongkhraam. 1983. Microbiological analysis of Thai fermented foods. *Journal of the Graduate School, Chulalongkorn University* 4: 11–24.

Wei, D. and S. Jong. 1983. Chinese rice pudding fermentation: Fungal flora of starter cultures and biochemical changes during fermentation. *Journal of Fermentation Technology* 61 (6): 573–579.

Welthagen, J. J. and B. C. Viljoen. 1998. Yeast profile in Gouda cheese during processing and ripening. *International Food Microbiology* 41 (3): 184–359.

Westall, S. and O. Filtenborg. 1998. Yeast occurrence in Danish Feta cheese. *Food Microbiology* 15 (2): 215–222.

Wood, B. J. B. and W. H. Holzapfel. 1995. *The lactic acid bacteria*. Vol. 2, *The genera of lactic acid bacteria*. London: Blackie Academic and Professional.

Wu, Y. C., B. Kimura, and T. Fujii. 1999. Fate of selected food-borne pathogens during the fermentation of squid *shiokara*. *Journal of Food Hygienic Society of Japan* 40 (3): 206–210.

Wu, Y. C., B. Kimura, and T. Fujii. 2000. Comparison of three culture methods for the identification of *Micrococcus* and *Staphylococcus* in fermented squid shiokara. *Fisheries Science* 66: 142–146.

Yang, R. and B. Ray. 1994. Factors influencing production of bacteriocins by lactic acid bacteria. *Food Microbiology* 11 (4): 281–291.

Yegna Narayan Aiyar, A. K. 1953. Dairying in ancient India. *Indian Dairyman* 5: 77–83.

Yokotsuka, T. 1985. Fermented protein foods in the Orient, with emphasis on shoyu and miso in Japan. In *Microbiology of fermented foods*. Vol. 1, ed. B. B. Wood, 197–247. London: Elsevier Applied Sciences.

Yokotsuka, T. 1991. Nonproteinaceous fermented foods and beverages produced with koji molds. In *Handbook of applied mycology*. Vol. 3, ed. D. K. Arora, K. G. Mukerji, and E. H. Marth, 293–328. New York: Marcel Dekker.

Yong, F. M. and B. J. B. Wood. 1977. Biochemical changes in experimental soy sauce *koji*. *Journal of Food Technology* 12: 163–175.

Yonzan, Y. 2007. Studies on selroti, a traditional fermented rice product of the Sikkim Himalaya: Microbiological and biochemical aspects. Ph.D. thesis, Food Microbiology Laboratory, Sikkim Government College, North Bengal University.

Yonzan, H. and J. P. Tamang. 1998. Consumption pattern of traditional fermented foods in the Sikkim Himalaya. *Journal of Hill Research* 11 (1): 112–115.

Yonzan, H. and J. P. Tamang. 2009. Traditional processing of *Selroti*—a cereal-based ethnic fermented food of the *Nepalis*. *Indian Journal of Traditional Knowledge* 8 (1): 110–114.

Yoon, S. S. 1993. *Che Min Yo Sul, A Translation of Chi-Min-Yao-Shu in Korean*. Seoul: Min Eum Sa (Korean).

Yoshida, S. 1993. Daizu Hakkou Shokuhin no Kigen. In *Nihon Bunka no Kigen: Minzokungaku to Idengaku no Taiwa*, ed. K. Sasaki and K. Mori, 229–256. Tokyo: Kodansho (Japanese).

Zhang, S. and Y. Liu. 2000. *Prepare technology of seasoning in China*. Beijing: South China University of Technology Press (Chinese).

Zvauya, R., T. Mygohi, and S. Parawira. 1997. Microbial and biochemical changes occurring during production of masvusru and mangisi, traditional Zimbabwean beverages. *Plant Foods Human Nutrition* 51: 43–51.

Index